North American Mixing Forum 2019

AA009164

Held at the 2019 AIChE Annual Meeting

Orlando, Florida, USA
10-15 November 2019

ISBN: 978-1-7138-0530-4

Printed from e-media with permission by:

Curran Associates, Inc.
57 Morehouse Lane
Red Hook, NY 12571

Some format issues inherent in the e-media version may also appear in this print version.

Printed with permission by Curran Associates, Inc. (2020)

For permission requests, please contact AIChE
at the address below.

AIChE
120 Wall Street, FL 23
New York, NY 10005-4020

Phone: (800) 242-4363
Fax: (203) 775-5177

www.aiche.org

Additional copies of this publication are available from:

Curran Associates, Inc.
57 Morehouse Lane
Red Hook, NY 12571 USA
Phone: 845-758-0400
Fax: 845-758-2633
Email: curran@proceedings.com
Web: www.proceedings.com

TABLE OF CONTENTS

557708 Dispersion and Coalescence in Liquid/Liquid Systems Stabilized with Nanoparticles

Nanoparticles can be applied as emulsifying agents to stabilize liquid/liquid systems towards coalescence (Pickering emulsions). Due to the adsorption of nanoparticles at the liquid/liquid interface, coalescence is hindered or arrested and smaller drop size distributions and a higher interfacial area can be achieved. Since this promotes mass transfer, nanoparticles can be used as innovative additives in liquid/liquid reaction systems. In this work, physical properties such as rheology and interfacial phenomena are investigated experimentally. Furthermore, the impact of different silica nanoparticles on drop size distributions in liquid/liquid systems is determined using microscopy or an in-situ endoscope technique. Thereby, emulsions generated with two different devices are compared: 1) High-energy devices such as ultrasonication, leading to emulsions that are nearly completely stabilized towards coalescence. 2) Low energy input induced by stirring, leading to only partly stabilized emulsions. This allows quantifying drop breakage and coalescence phenomena in steady state and dynamic conditions. The experiments can provide essential information for model development, e.g., breakage and coalescence rates used in population balance equations.

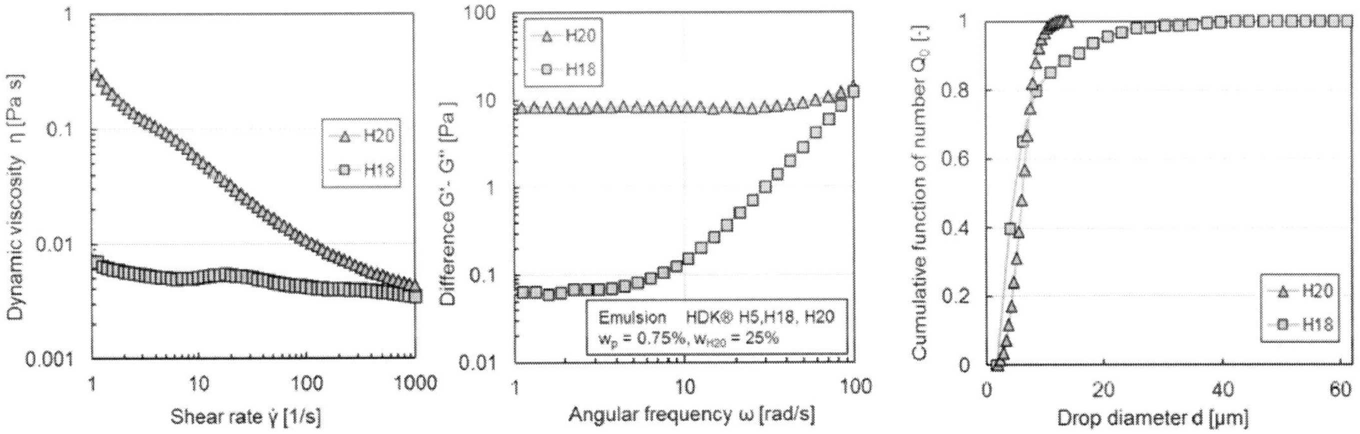

Figure 1: Apparent emulsion rheology, frequency sweep results and drop size distributions in w/o emulsions stabilized by two silica nanoparticles (HDK H18 and H20) that differ in wettability. Emulsions were prepared with ultrasonication and are stable towards coalescence. (Weight percent of water w_{H2O} = 25%, oil phase = 1-dodecene, particle weight percent wp = 0.75%) [1]

The presence of nanoparticles in suspensions and emulsions leads to complex rheological behaviour. The systems become shear thinning and viscoelastic with gel-like character especially at high particle concentrations and small droplet sizes. The nanoparticles are initially suspended in one liquid phase and move to the liquid/liquid interface as the systems are dispersed and droplets are created. The free nanoparticles, the nanoparticles at the interface and the droplet sizes influence the rheology of the system. Comparing silica nanoparticles with different wettability (HDK H18 and H20) in completely stabilized w/o emulsions leads to significant changes in emulsion rheology (Figure 1 left, middle). The nanoparticle type affects the particle-particle and particle-fluid interactions, so that the rheology also depends on particle morphology, size and concentration. Furthermore, the drop size distribution of the system also is shifted towards smaller diameters for the particles with intermediate wettability HDK H20 in comparison to the more hydrophobic HDK H18 [1].

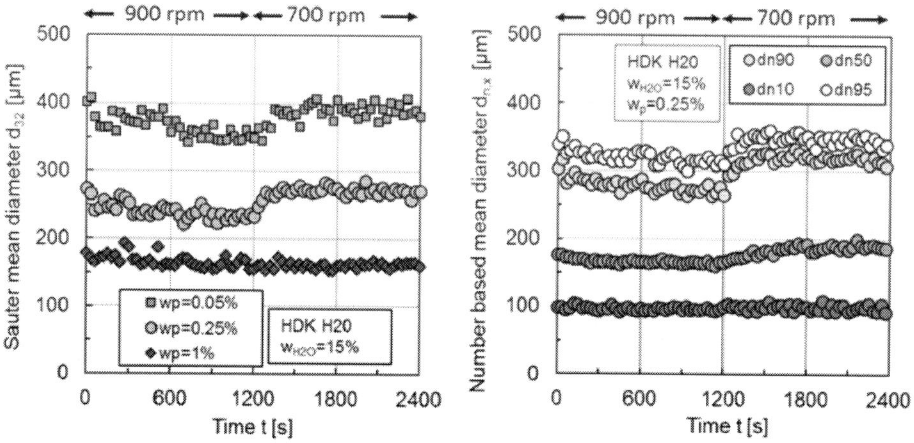

Figure 2: Dynamic changes in Sauter mean diameter (left) and other characteristic diameters (right) after abrupt changes in energy input in a w/o emulsion dispersed in a stirred tank. (Weight percent of water w_{H20} = 15%, oil phase = n-heptane, particle weight percent wp = 0.05 – 1.00%) [2].

The aforementioned parameters also influence the drop sizes of the emulsions in systems with low energy input that are only partly stabilized towards coalescence. Figure 1 (left) shows transient Sauter mean diameters for different particle concentrations after agitation start and after an abrupt change of agitation speed from 900 to 700 rpm. The effect of coalescence inhibition for different particle concentrations is clearly visible. With rising particle content, the Sauter mean diameters barely increase after energy input reduction. Figure 1 (right) depicts other characteristic diameters for one particle concentration and indicates that the lowest drop fraction is stabilized against coalescence, whereas the larger drops still coalesce after reduction in energy input. This shows that the shape of the drop size distribution is affected [2].

[1] Hohl, L.; Röhl, S.; Stehl, D.; von Klitzing, R.; Kraume M.: Influence of nanoparticles and drop size distributions on the rheology of w/o Pickering emulsions. Chem. Ing. Tech. 88, No. 11., 1818-1826

[2] Röhl, S.; Hohl, L.; Kempin, M.; Enders, F.; Jurtz, N.; Kraume, M.: Influence of different silica nanoparticles on drop size distributions in agitated liquid/liquid system. submitted to Chem. Ing. Tech. in 2019

571812 Thixotropic Suspensions in Static Mixer: Comparison of Different Structure-Based Models

Thixotropy is one of the most challenging problems in colloid rheology. A comprehensive definition for thixotropy proposed by Mewis and Wagner (2009) as "...the continuous decrease of viscosity with time when flow is applied to a sample that has been previously at rest, and the subsequent recovery of viscosity when the flow is discontinued". This behavior of the working fluid is very common in industrial applications. Thixotropy can be encountered in many industrial applications, including the processing of pharmaceuticals, minerals, food products, coatings, paints, gels, inks, drilling muds, concrete, etc. In this work, we investigate thixotropic behavior of the fluid in a static mixer by using numerical techniques.

In this work, using SIMCENTER STAR-CCM+ a continuum approach based on structure variable transport equation was adopted to account structure buildup and breakdown in the fluid. We used different structure models proposed by Mujumbar et al (2002), Houska (1980), and Coussot et al (2002), to analyze mixing pattern and viscosity decay in static mixer. Furthermore, irreversible structure model also investigated, and the results were compared with proposed reversible material models. Our numerical approach was fully verified by comparing the fluid flow properties and structure variable results with those reported in the literature.

Mewis, J., & Wagner, N. J. (2009). Thixotropy. Advances in Colloid and Interface Science, 147, 214-227.

Mujumdar, A., Beris, A. N., & Metzner, A. B. (2002). Transient phenomena in thixotropic systems. Journal of Non-Newtonian Fluid Mechanics, 102(2), 157-178.

Houska, M. Inzenyrske aspekty reologie tixotropnich kapalin. Diss. Ph. D. thesis, Technical University of Prague, Praha, Czechia, 1980.

Coussot, P., Raynaud, J. S., Bertrand, F., Moucheront, P., Guilbaud, J. P., Huynh, H. T., & Lesueur, D. (2002). Coexistence of liquid and solid phases in flowing soft-glassy materials. Physical review letters, 88(21), 218301.

573374 Using Thrust As an Indirect Measure for Mixing Efficiency of Submerged Propeller Agitators in Non-Newtonian and Viscoelastic Media

Using thrust as an indirect measure for mixing efficiency of submerged propeller agitators in non-Newtonian and viscoelastic media

Markus Kolano, Manuel Brehmer, Matthias Kraume, TU Berlin, Chair of Chemical and Process Engineering, Ackerstraße 76, 13355 Berlin, Germany

In Germany, most biogas fermenters are mixed using one or multiple submerged propeller agitators. The composition of fermented substrates is highly variable and changes in rheological behavior can lead to an insufficient mixing performance. Thus, operational parameters like stirring rates as well as geometrical orientation of the agitators have to be continuously adjusted to the rheological properties. Evaluating the efficiency of these mixing systems is difficult, since used substrates are opaque and optical measurement techniques are not applicable. Indirect measures are therefore needed.

In this talk, we evaluate whether thrust can be used to determine the mixing efficiency of aforementioned systems. Using the particle image velocimetry method in combination with thrust measurements via strain gauges as well as mixing times determined by decolorization, relationships between flow numbers, axial forces and mixing times are experimentally derived for various single and multiple propeller mixing systems in laboratory scale. For that, transparent model fluids with varying (viscoelastic) flow properties are used, allowing for the derivation of rheological dependencies. The experimental data also serves as validation basis for numerical simulations. In these, viscoelastic flow properties of the model fluids as well as bio substrates, which are determined using a novel relative measurement technique, are used.

The general flow behavior of the propeller jets can be categorized by the elasticity number (Fig. 1, left). While in the elastically dominated regime an axial flow profile is produced, the interplay of inertial and elastic forces causes distinct radial flow components in the transition regime. In that regime, the axial flow component of the propeller jet can be well described by thrust (Fig. 1, right). Since mixing times are predominantly dependent on axial flow in the here analyzed mixing systems, we can show thrust can finally be used to estimate mixing times as well.

Figure 1: Propeller jet angle in dependence of elasticity number (left); axial flow and thrust in the transition regime (right)

563887 Macromixing Characteristics of Viscous and Shear-Thinning Fluids in Rotor-Stator Spinning Disc Reactors

Macromixing Characteristics of Viscous and Shear-Thinning Fluids in Rotor-Stator Spinning Disc Reactors

Arnab Chaudhuri[a] (a.chaudhuri@tue.nl), Wyatt Winkenwerder[b] (Wyatt.Winkenwerder@nouryon.com), John van der Schaaf[a] (J.Vanderschaaf@tue.nl)

Department of Chemical Engineering, TU Eindhoven, 5612 AZ, Eindhoven, Netherlands

Nouryon-Surface Chemistry, Croton River Center 281 Fields Lane, Brewster 10509, New York, United States of America

Introduction

In recent years the chemical industry has been focused on improving both sustainability and safety of chemical processes. This involves for instance an increase in efficiency of raw material/energy usage. This drive towards process intensification has seen the development of many novel, integrated reactors one of which is the rotor-stator spinning disc reactor (rs-SDR). The rs-SDR consists of a rotating disc, enclosed by a stationary cylindrical housing. Intensification in mass transfer and heat transfer rates can be achieved due to the small cavities which provide a high degree of turbulence and shear. Previous investigations into the hydrodynamics of rs-SDRs have resulted in an engineering model to describe the macromixing characteristics for aqueous systems as a function of flow rate, gap distance and rotation speed[1]. As we look towards applications for the rs-SDR, it becomes vital to characterize viscous systems, both Newtonian and non-Newtonian, as these are highly relevant for industrial systems. In this study, we have characterized the macromixing behavior of viscous systems using residence time distribution measurements and have looked at how they differ in comparison with aqueous systems at varying rotation speeds.

Experimental Methods

A single stage rs-SDR has been used for this study. The radius of the rotor is $r_d = 143*10^{-3}$ m while the radius of the cylindrical housing is $145*10^{-3}$ m. This results in a radial gap of $2*10^{-3}$ m. The height of the stage is $9*10^{-3}$ m with a disc thickness of $4*10^{-3}$ m, therefore the resulting axial gap is $2.5*10^{-3}$ m.

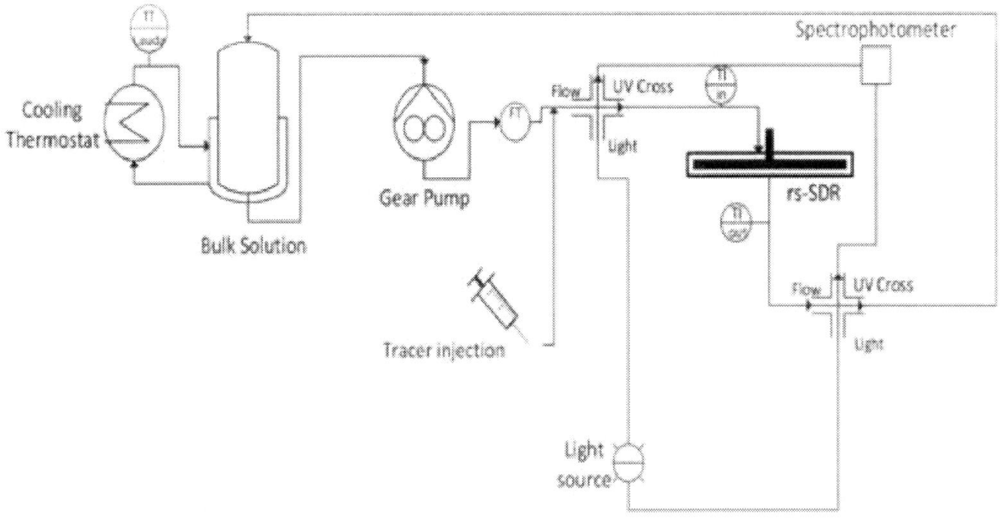

Figure 1: Schematic of the experimental setup.

The liquid was pumped at the top of the reactor with flowrates monitored by a CORIFLOW^{TM} (Bronkhorst) meter and was withdrawn from the bottom. The temperature in the system is maintained with the use of a cooling bath (LAUDA ECO RE 630). The RTD measurements were carried out by measuring the tracer injection through UV-vis at both the inlet and outlet. By deconvolution of the inlet signal from the outlet we were able to obtain the final RTD curve[2]. A schematic of the experimental setup is shown in Fig. 1. We have used glycerol and Xanthan gum (1 w/w%) to study the behavior of viscous and non-Newtonian fluids, respectively.

Results & Discussion

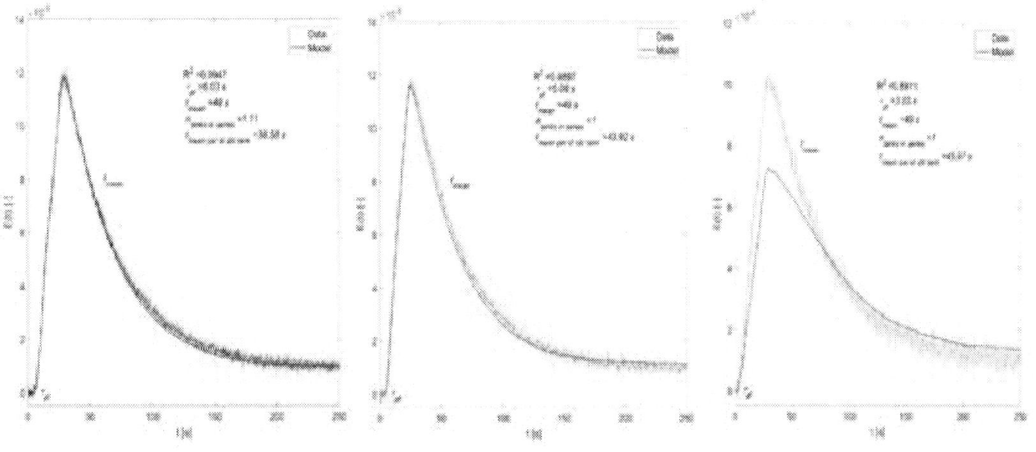

(A) (B) (C)

Figure 2: Experimentally obtained E_{exp} curves with the conventional PFR-CSTR in series model fitting (G=0.035, Re_{ω}= 4.27 * 10^5, C_{ω} = 104.7). A) Water B) Glycerol (69%w/w) C) Xanthan Gum (0. 1% w/w)

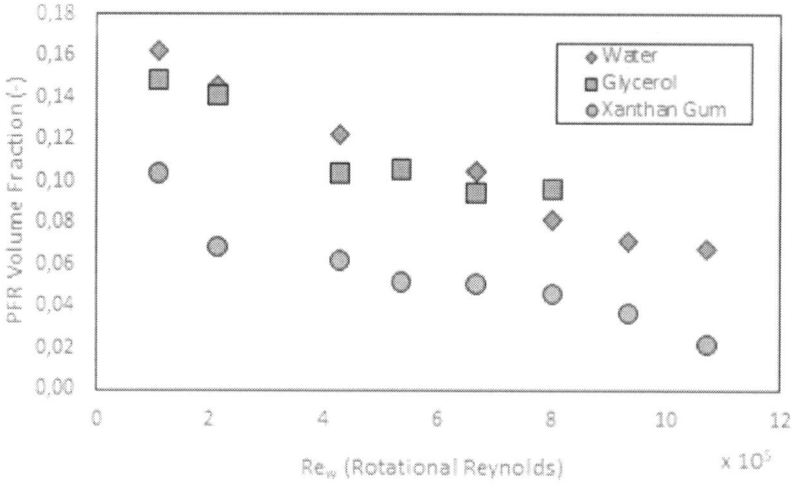

Figure 3: Effect of rotation speed on the PFR volume fraction for various fluids in the RS-SDR

The results indicate that while the model used for aqueous systems was applicable to viscous, Newtonian systems; for Xanthan gum (non-Newtonian) we found that the fitting of the E-curve was quite poor (Fig 3C). This is most likely due to the presence of varying regions of viscosity in the reactor. At the edge of the disc where the shear is most intense, we can expect that the viscosity to be low, while near the entrance and exit of the fluid stream we can expect the viscosity to be much higher. Therefore, an alternative model should be investigated for this system, mostly likely involving CSTRs with exchange volumes to differentiate between regions of higher and lower viscosity.

We have also observed that the time at which we first observe a signal at the outlet also varies significantly for the non-Newtonian fluid. For the aqueous model, this breakthrough time was described as the PFR residence time. Assuming that PFR volumes are still present in the reactor for non-Newtonian fluids, we can see that the PFR volume of the reactor decreases significantly for Xanthan Gum in comparison to Glycerol or Water (Fig 3). This can be expected since the shear thinning nature of the fluid will mostly likely lead it to transition to CSTR type behavior at earlier reactor lengths.

Future Work

The results obtained thus far have illustrated the applicability of RS-SDRs for viscous flows. We have observed that the macromixing characteristics of the reactor can still be maintained for Newtonian fluids with kinematic viscosities greater than 27 times that of aqueous systems. We are currently investigating at what values the model for the viscous system is unable to mimic the aqueous system and poor mixing behavior is observed. We are also investigating a model to describe the non-Newtonian fluid in a RS-SDR. Furthermore, this work will be extended to also investigate the effect of flow rate and gap ratio on the macromixing behavior of viscous flows.

References

[1] M.M. de Beer, J.T.F. Keurentjes, J.C. Schouten, J. van der Schaaf, Engineering model for single-phase flow in a

multi-stage rotor–stator spinning disc reactor, Chem. Eng. J. 242 (2014) 53–61. doi:10.1016/j.cej.2013.12.052.

[2] D. Boškovic, S. Loebbecke, G.A. Gross, J.M. Koehler, Residence Time Distribution Studies in Microfluidic Mixing Structures, Chem. Eng. Technol. 34 (2011) 361–370. doi:10.1002/ceat.201000352.

569975 Computational Fluid Dynamics Modeling of Mixing and Reaction of High-Viscosity Liquids in a Continuous Flow Reactor

Process intensification is a strong trend in chemical manufacturing which aims to conduct chemical processes with drastically increased energy and resource efficiency, lower capital and/or operating cost, and reduced physical and environmental footprint. In this context, the transition from traditional batch to continuous processes, which has been conducted in the pharmaceutical industry over the past decade, is now reaching the much larger scale of specialties and fine chemicals production. Among the many challenges faced during this transition is the requirement for rapid mixing of reactants in order to achieve optimum conversion and selectivity and consistent product quality. This problem is more pronounced in cases where the reactants are immiscible and/or have high viscosities at operating temperatures, as is often the case for specialty chemicals such as lubricants and dispersants.

It is well known that mixing is affected by material properties such as density, viscosity and surface tension; by operating conditions such as pressure and temperature; and by flow conditions and geometries. Chemical reactions between the mixing liquids further affects the mixing patterns in the reactor by varying the properties and local temperatures at different points in the fluid. However, while mixing has been studied for many decades, the detailed mixing behavior of such complex systems is still not fully understood, yet critical for reactor design and operation.

In the present study, we present computational fluid dynamics (CFD) simulations aimed towards understanding the mixing behavior of highly viscous liquids (polymeric anhydrides and amines) that are typically used in manufacturing dispersants via amination of the anhydride. Beyond a systematic analysis of the mixing behavior inside a continuous flow tubular reactor as a function of some of the variables listed above, we will discuss the use of appropriate metrics to study mixing, such as coefficient of variation, exposure and reaction conversion. Furthermore, the effect of different inlet geometries—ranging from co-axial flows to spray-type feed configurations—as well as the use of in-line static mixers will be discussed. Key results from CFD simulations are validated against laboratory experiments with non-reactive flow visualization as well as reactive tubular reactor experiments. The results of this study are being used to assist optimal design of experiments and scaling up to pilot-plant scale.

573750 Non-Invasive Detection and Assessment of Coronary Stenosis Using Mean Age of Blood Flow

We developed a new method and metric to non-invasively detect and assess coronary stenosis (blocked arteries) based on mean age of blood flow through diseased arterial segments. The mean age of blood was first computed using CFD in model coronary arteries with varying degrees of stenosis and then in 100 patients who underwent the invasive FFR procedure accepted as the gold standard of clinical diagnosis. The CFD model was validated by predicting mean age within 1.3% of the nominal mean flow time (volume divided by flow rate) in normal arteries. A dimensionless form of the age metric accounts for differences in segment volume and flow rate, and is indicative of the increase in age as compared to the nominal age for a given arterial segment. Blood age in diseased arteries increased as much as 60% relative to the expected nominal mean age, and the increase is due to flow characteristics as blood exits the constricted area, specifically small regions of recirculatory flow just past the stenosis. Acceleration and deceleration of blood due to the pulsing flow contributes to holdup in this region. Our blood age metric correlated strongly with the invasive FFR metric, withAUC, specificity, sensitivity all greater than 90%. More importantly, our metric agreed with the decision of the invasive FFR metric on whether or not to intervene with a stent in 98/100 cases. When applied to clinical practice, this could potentially allow practicing cardiologists to accurately assess and quantify the severity of coronary stenosis without resorting to invasive catheter-based techniques.

579590 Further Studies on Optimization of Agitation in Suspension Polymerizers

Suspension polymerization is a common process for making products such as PVC. Droplets of monomer are dispersed in water in the presence of a surfactant and once the desired droplet size is achieved an initiator is added to start the reaction. The size of the polymer particles produced is governed by the size of the monomer droplets formed at the start of the process.

As the reaction proceeds, coagulum is formed which can coat the impellers and vessel walls or form large masses which can damage the agitator if they impact the blades or shaft. The formation of coagulum can be minimized by using an impeller that generates a high flow but, at the same time, it must provide sufficient shear to generate the desired monomer droplets.

Droplet sizes in stabilized liquid-liquid dispersions have been measured to identify which impellers can be characterized as "high or low shear". The droplet sizes are correlated in terms of the kinetic energy dissipation rate (KEDR) in the impellers' trailing vortices (see Grenville et al., Chem. Engg., Aug 2017). The measured relationship between the impellers' characteristics is used to validate prediction of the trailing vortex KEDR made with large-eddy simulation computational fluid dynamics.

A case study on retrofitting in a suspension polymerization reactor will be presented.

573834 A Novel Approach to the Experimental Determination of N_{js} Using Electrical Resistance Tomography

N_{js}, the minimum agitation speed needed to just suspend all the solid particles in a solid-liquid mixture stirred in an agitated vessel, is a critical parameter to properly operate industrial mixing tanks and reactors in a large number of industrial operations. As a result, a significant literature on N_{js} is available. The oldest and most common method to measure N_{js} experimentally is that of Zwietering's (*Chem. Eng. Sci.*, 1958, 8, 244-253), where N_{js} can be visually obtained by determining the agitation speed at which the solids remain at the bottom of the tank for no more than 1-2 seconds before being swept away. Although this has been shown to be a reliable method to obtain N_{js}, it still relies on visual observation of the bottom of the vessel and it is therefore potentially susceptible to observer's bias. To address this issue new observed-independent experimental approaches to determine N_{js} using measurements of the fraction of solids on the vessel bottom were previously developed by our research group. However even those methods are unsuitable to be used in opaque fluids/tanks or if images of the vessel bottom cannot be taken.

In order to experimentally determine N_{js} even in systems where the tank content cannot be inspected, a novel method using Electrical Resistance Tomography (ERT) was developed and tested in this work. Accordingly, a sensor array probe consisting of a straight plastic rod mounting 16 electrodes was placed vertically in a fully baffled mixing tank provided with a centrally mounted agitation system and containing water and non-conductive glass beads approximately 300 μm in diameter. The electrodes were connected to an external ERT system and data acquisition apparatus (P2+ System, Industrial Tomography Systems, Manchester, UK) dynamically measuring the conductivity distribution and resistivity in the solid-liquid system data between each pair of adjacent electrodes. The apparatus consisted of a signal source, voltmeters, electrode multiplexer array, signal demodulators, and a system controller, connected to a computer. The system generated Alternating Current (AC) between pairs of neighboring electrodes and the resulting voltage was measured across all other neighboring electrodes. Current injection was applied to all neighboring electrodes. Through this approach it was possible to measure the mean bulk resistance across the electrodes on the sensing array probe, measure the conductivity distribution on a portion of a vertical place inside the tank, and also obtain 2-D images of the conductivity distribution inside the tank through the use of image reconstruction algorithms.

In a typical experiment the array probe was placed in the tank, and after proper calibration, the mean bulk resistance of the solid-liquid mixture was obtained as the mixture was stirred by an impeller in the mixing tank at different values of the impeller agitation speed, N. As N increased, increasingly larger fractions of the non-conducting solids became suspended, thus increasing the resistivity of the suspension measured by the ERT apparatus. A plot of the percent resistance variation vs. the agitation speed resulted in an *S*-shaped curve, which eventually reached an asymptotic limit value as all solids became suspended and dispersed in the liquid. In order to extract N_{js} from the data, a mathematical approach previously developed by our groups for a different system was used (Huang and Armenante, *Chem. Eng. Sci.*, 1992, 47, 2865-2870). Accordingly, the experimental data were interpolated with cubic spine curves and the agitation speed at which the function F(N), equal to the ratio of the second derivative to the first derivative of the combined spine curve function, showed a minimum point was takes as the N_{js} value ($N_{js\text{-}ERT}$). The rationale for this approach is as follows: F(N) represents how the change in slope of the spline curve (second derivative) with respect to the spline curve slope (first derivative) varies with N. F(N) can be expected to reach a minimum value when the spline function is just about to bend to approach the asymptote.

Experiments were conducted where $N_{js\text{-}ERT}$ was obtained under different operating conditions, i.e., where the impeller type, impeller ratio-to-tank diameter ratio, and impeller clearance were varied. N_{js} was not only experimentally obtained using the proposed ERT approach but also using the Zwietering method as well as the method recently developed by Shastry and Armenante (AIChE Meeting, Minneapolis, 2017).

Then, parity plots were constructed in which $N_{js\text{-}ERT}$ was plotted against the N_{js} values obtained with the other two methods. Excellent agreement was observed in all plots, indicating that the novel method proposed here can be effectively used for the experimental determination of N_{js}.

The results of this work show that ERT combined with the analysis of the data proposed here can be used to effectively measure N_{js} in solid-liquid dispersion in mechanically stirred vessels. The proposed approach is observer-independent and can be used even in systems that cannot be directly observed, such as industrial tanks. Therefore, it is expected that this approach could find extensive practical applications in the chemical, pharmaceutical and biopharmaceutical industries.

572575 Characterization of Micromixing in a Rotor-Stator Spinning Disc Reactor: Effects of Gap Distance

In the chemical industry many processes involve complex reaction systems where fast reactions take place, deriving in a range of products while in fact only one is the desired one. In the past, in order to achieve high selectivity and yield towards the main product, a common approach has been to slow down the reactions by diluting the feeding streams, usually with solvents. As a consequence, large reactors are needed to meet the production requirements, and the removal of the solvents derive in high energy costs.

More recently, novel equipment has been developed in the framework of Process Intensification (PI), aiming to enable production conditions that were not possible with traditional equipment by the enhancement of mass and heat transfer rates. With these novel reactors, high yields can be obtained without the need to dilute the reaction systems.

In those high concentrated systems with higher reaction rates, the selectivity is then determined by the mixing efficiency, especially in systems where competitive consecutive or parallel competitive reactions take place. More specifically, for very fast competitive reactions the micromixing time -which is the homogenization of the system at the molecular scale- determines the product distribution. The micromixing time has been extensively investigated and well correlated to the turbulence intensity, mainly in stirred vessels where homogenous and isotropic turbulence is observed.

For novel reactors such as the rotor-stator spinning disc reactor, high shear forces applied to very small reaction volumes lead to a very high turbulence intensity. These are in the range of 4 to 6 orders of magnitude higher than those achieved in stirred vessels. In other words, very high micromixing efficiency is expected[1].

In literature, there are models that can describe the relationship between micromixing time and energy dissipation rate. The micromixing time has been theoretically and experimentally determined to be proportional to the Kolmogorov microscale:

$$t_{mix} = C\left(\frac{\nu}{\varepsilon}\right)^{0.5} \qquad (1)$$

Where v is the kinematic viscosity of the fluid, e is the (local) energy dissipation rate. One of the most consistent models based on turbulence theory is the Engulfment model[2]. In this model the term C takes a value of 17.23, based on a derivation from the most hydrodynamic eddy of the turbulent field engulfing liquids from the surroundings and allowing a small reaction zone to grow and react. Similarly, other models like the incorporation model[3], uses this term C as a correlation to obtain a micromixing time that is proportional to the incorporation of liquids from the surroundings of the reaction zone.

Furthermore, some test reaction systems have been developed, in order to determine the micromixing times[4][3]. Assuming enough knowledge of the kinetics of the system, the fundamental basis of these test reactions is that by performing competitive reactions, one can analyze the product distribution and relate it to a micromixing time, using one of the available models.

In the rotor-stator spinning disc reactors, the shear rate and therefore the energy dissipation rate is known to have a proportionality to the radius of the disc[5][6]:

$$\varepsilon \propto r^{11/4} \qquad (2)$$

The previous relationship allows for a good estimation of the local energy dissipation rate by performing a momentum balance.

Naturally, by reducing the gap distance between the rotor and the stator, the reactor's volume will also decrease. This suggests that the energy transferred from the disc into the liquid will be locally higher, and also a function of the gap clearance. In the turbulent regime with merged boundary layers, the shear rate – and energy dissipation rate– is a function of the gap distance [5][6]:

$$\varepsilon \propto h^{-1/4} \qquad (3)$$

Experimental results in a rotor-stator spinning disc reactor at various gap distances show, however, that there is not significant effect observed when modifying this parameter. In the first graph it can be observed that the estimated local energy dissipation rate obtained from the momentum balance is underestimated for bigger gap distances. On the other hand, on the second graph the results indicate that the main parameter to define the micromixing time is the rotational speeds.

The results suggest that the average local energy dissipation rate estimated from the correlations from literature [6] give a good approximation for the macromixing effects (bubble formation[5], convective heat transfer[7], mass transfer [8]) in the enclosed space between a rotating disc and a wall . However, for the non-homogeneous, anisotropic turbulent field created by the high shear forces in such a small volume, the local energy dissipation rate requires a correlation to the tangential velocity. This is a topic of ongoing research.

Furthermore, a comparison between two test reaction systems is presented, adopting a new kinetic model and addressing previous concerns about the use of the Villermaux-Dushman system (unpublished work). With this new kinetic model, lower micromixing times are obtained than those previously reported using the Villermaux-Dushman method when compared to the Diazo-coupling of naphthols [9]. This is however not surprising, since the iodide-iodate reaction in the Villermaux-Dushman is much faster (kinetic coefficient in the order of magnitude 8) than the diazo coupling of 2-naphthol (kinetic constant in the order of magnitude 5). The iodide-iodate reaction occurs in a more localized zone with higher turbulent intensity.

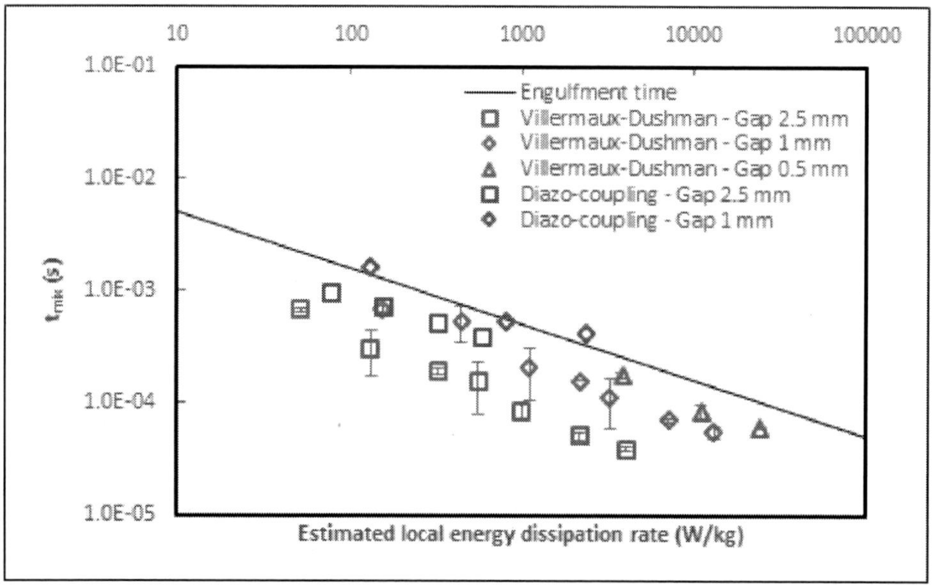

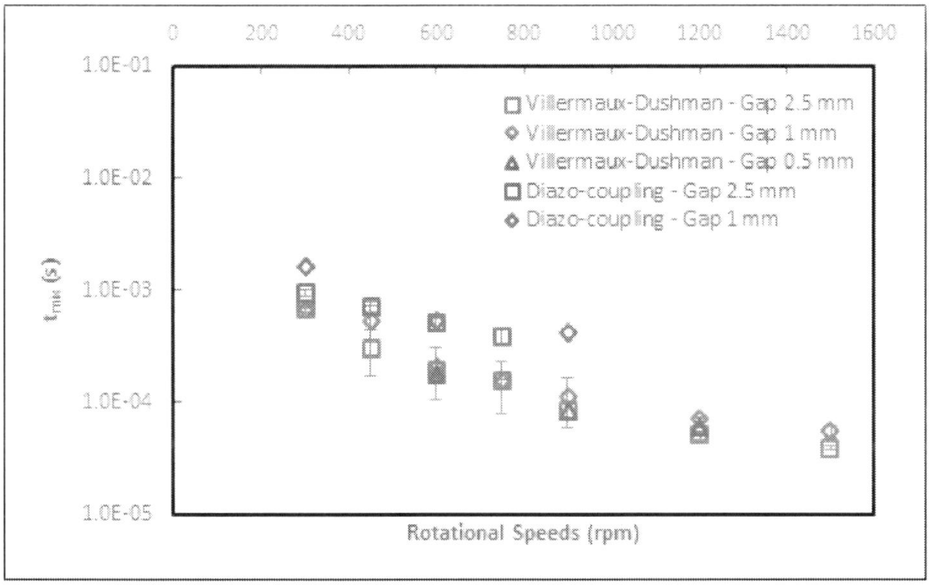

[1] A. N. Manzano Martínez, K. M. P. Van Eeten, J. C. Schouten, and J. Van Der Schaaf, "Micromixing in a Rotor-Stator Spinning Disc Reactor," *Ind. Eng. Chem. Res.*, vol. 56, no. 45, pp. 13454–13460, Nov. 2017.

[2] J. Baldyga and J. R. Bourne, "Simplification of micromixing calculations. I. Derivation and application of new model," *Chem. Eng. J.*, vol. 42, no. 2, pp. 83–92, Nov. 1989.

[3] M.-C. Fournier, L. Falk, and J. Villermaux, "A new parallel competing reaction system for assessing micromixing efficiency—Determination of micromixing time by a simple mixing model," *Chem. Eng. Sci.*, vol. 51, no. 23, pp. 5187–5192, Dec. 1996.

[4] J. R. Bourne, O. M. Kut, and J. Lenzner, "An improved reaction system to investigate micromixing in high-intensity mixers," *Ind. Eng. Chem. Res.*, vol. 31, no. 3, pp. 949–958, Mar. 1992.

[5] M. M. de Beer, J. T. F. Keurentjes, J. C. Schouten, and J. van der Schaaf, "Bubble formation in co-fed gas–liquid flows in a rotor-stator spinning disc reactor," *Int. J. Multiph. Flow*, vol. 83, pp. 142–152, Jul. 2016.

[6] M. Djaoui, A. Dyment, and R. Debuchy, "Heat transfer in a rotor-stator system with a radial inflow," *Eur. J. Mech. B/Fluids*, vol. 20, no. 3, pp. 371–398, 2001.

[7] M. M. de Beer, J. T. F. Keurentjes, J. C. Schouten, and J. van der Schaaf, "Convective condensation in a stator-rotor-stator spinning disc reactor," *AIChE J.*, vol. 62, no. 10, pp. 3784–3796, Oct. 2016.

[8] M. Meeuwse, J. van der Schaaf, B. F. M. Kuster, and J. C. Schouten, "Gas–liquid mass transfer in a rotor–stator spinning disc reactor," *Chem. Eng. Sci.*, vol. 65, no. 1, pp. 466–471, Jan. 2010.

[9] P. Guichardon, L. Falk, and M. Andrieu, "Experimental Comparison of the Iodide-Iodate and the Diazo Coupling Micromixing Test Reactions in Stirred Reactors," *Chem. Eng. Res. Des.*, vol. 79, no. 8, pp. 906–914, Nov. 2001.

571372 (Micro-)Mixing Effect in Alkylbenzene Sulfonation Using Spinning Disc Technology

Since the introduction of process intensification numerous chemical processes have been successfully optimized and have obtained an excellent control on conversion and selectivity, without making any concessions on process safety and production rate[1]-[3]. The idea of working at the intrinsic kinetic rate rather than having mass or heat transfer limitations gave rise to intensified compact reactors which have the potential to achieve the same throughput as conventional large batch reactors. Such compact reactors include micro reactors, rotating packed beds, and spinning disc reactors (SDR). For example, a significant decrease in heat[4], [5] and mass[6]-[8] transfer limitations were reported for the SDR. Furthermore, the local mixing time was measured to be as low as 1.13×10^{-4} seconds[9].

Technologies such as the SDR are especially useful when working with reaction systems which are highly exothermic, are very rapid in terms of reaction rates, consist of multiple phases, and are very sensitive towards selectivity. An example is the extremely fast sulfonation of alkylbenzenes using fuming sulfuric acid to form the para-sulfonic acid. The formation of byproducts, such as unwanted ortho- and meta-sulfonic acid isomers as well as di sulfonic acids and sulfones, decreases the selectivity significantly.

In this work, a Synthetron Spinning Disc Reactor (S-SDR) was used to investigate the effects of mixing on the selectivity of the sulfonation of toluene and ethylbenzene reactions using fuming sulfuric acid. The reactor consists of a disc (a disc radius of 7.62 cm) rotating at a speed range of 50 to 9000 rpm inside a narrow cylindrical encasing with a gap width of 0.2 millimeter. Reactants were fed below the disc and the reaction mixture flowed upwards exiting the reactor around the disc shaft.

The experimental study was carried out by varying the rotational speed of the disc, while maintaining a constant residence time. This was tested for different molar flow ratios of the reactants. Several analysis techniques (C13-NMR, H-NMR, LC-MS and titration) were used to quantitatively determine the reaction products and their relative abundance. Additional analyses were conducted to determine the viscosity of the outflow as function of the composition. Figure 1 depicts the toluene and ethylbenzene results for different molar flow ratios. The graphs clarify the discussion below.

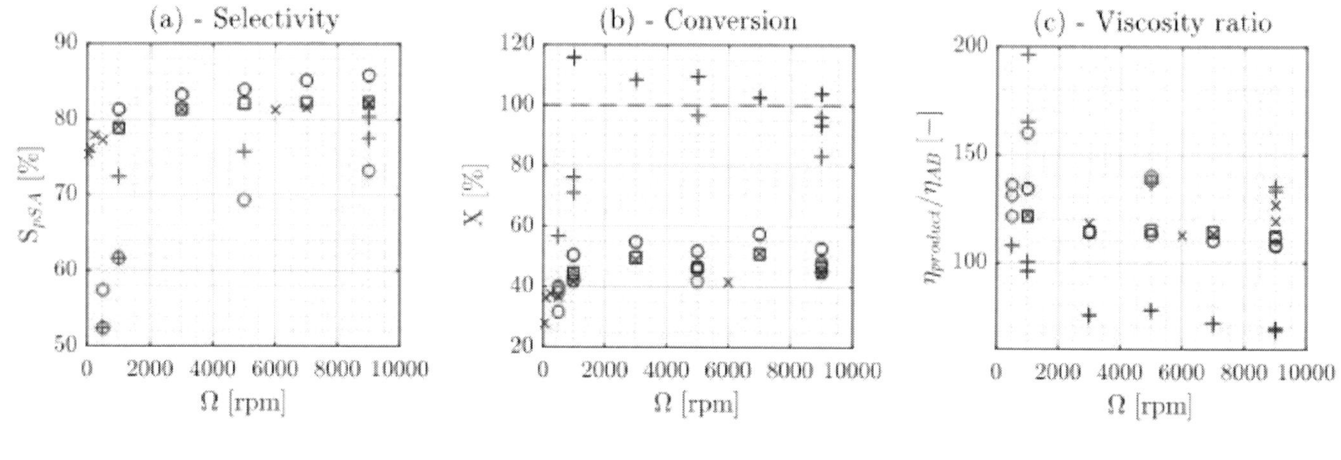

Figure 1 (a) Selectivity towards the para isomer (S_{pSA}), (b) conversion (X) and (c) viscosity ratio ($_{product}/_{AB}$) versus the rotational speed (O) for constant residence time (t). All measurements were done for different molar flow ratios (Ψ) which is defined as the inlet molar flowrate of the considered alkylbenzene over the inlet molar flowrate of fuming sulfuric acid. Toluene (red): + Ψ = 0.9, = 6.5 s ○ Ψ = 2.0, = 6.5 s. Ethylbenzene (blue): + Ψ = 0.8, = 6.4 s ○ Ψ = 1.6, = 4.3 s □ Ψ = 2.4, = 4.3 s × Ψ = 3.2, = 4.3 s

The results in figure 1 (a) show an increase in selectivity with increasing rotational speed (mixing intensity) whilst maintaining a constant conversion for both toluene and ethylbenzene (figure (b)). The selectivity of toluene sulfonation was predominantly determined by the formation of disulfonic acids and sulfones. For ethylbenzene sulfonation only the formation of disulfonic acids was found.

At constant conversion (i.e. 1000 rpm to 9000 rpm), the increase in selectivity was found to be 10% for the toluene experiments, while for the ethylbenzene experiments this was 3 to 5% depending on the different flow ratios. The latter showed that ethylbenzene sulfonation is the least sensitive towards mixing in line with the lower reaction rate as measured in earlier work[10]. Given the constant conversion, the observed shift in selectivity can only be explained in terms of local mixing phenomena rather than bulk effects such as heat or mass transfer. This is because bulk effects influence the overall reaction rate so that the conversion should change subsequently for a constant residence time. As was discussed above, this was not observed for both toluene and ethylbenzene sulfonation.

At low rotational speed the measured viscosity ratio ($_{product}/_{AB}$) peaked at circa 200 for the toluene case, while for the ethylbenzene this was 135 (figure 1 (c)). In both cases, the viscosity ratio decreased for increasing rotational speed. The lowest ratios at maximum rotational speed (9000 rpm) were 110 and 70 respectively. At this condition, maximum selectivity was obtained for all flow ratios. The strong increase in viscosity probably worsens the local turbulent mixing, resulting in locally formed hot-spots and large concentration gradients. This stresses the importance of local mixing in these systems.

[1] Flowid, "Butyl-Lithium Chemistry in the Flowid SpinPro Reactors," july 11th, 2017. [Online]. Available: https://www.flowid.nl/butyl-lithium-chemistry-the-flowid-spinpro-reactors/. [Accessed: 12-Dec-2018].

[2] S. Jovic, "Intensification of chlorine processing," Eindhoven: Technische Universiteit Eindhoven, 2016.

[3] P. Granados Mendoza et al., "Intensification of the chlor-alkali process by using a spinning disc membrane electrolyzer," Chem. Eng. Res. Des., vol. 128, pp. 120–129, Dec. 2017.

[4] M. M. de Beer, L. Pezzi Martins Loane, J. T. F. Keurentjes, J. C. Schouten, and J. van der Schaaf, "Single phase fluid-stator heat transfer in a rotor–stator spinning disc reactor," Chem. Eng. Sci., vol. 119, pp. 88–98, Nov. 2014.

[5] J. Kleiner, F. Haseidl, and O. Hinrichsen, "Rotor-Stator Spinning Disc Reactor: Characterization of the Single-Phase Stator-Side Heat Transfer," Chem. Eng. Technol., vol. 40, no. 11, pp. 2123–2133, Nov. 2017.

[6] M. Meeuwse, S. Lempers, J. van der Schaaf, and J. C. Schouten, "Liquid-Solid Mass Transfer and Reaction in a Rotor-Stator Spinning Disc Reactor," Ind. Eng. Chem. Res., vol. 49, no. 21, pp. 10751–10757, Nov. 2010.

[7] M. Meeuwse, J. Van Der Schaaf, B. F. M. Kuster, and J. C. Schouten, "Gas – liquid mass transfer in a rotor – stator spinning disc reactor," Chem. Eng. Sci., 2010.

[8] M. Meeuwse, J. van der Schaaf, and J. C. Schouten, "Mass Transfer in a Rotor-Stator Spinning Disk Reactor with Cofeeding of Gas and Liquid," Ind. Eng. Chem. Res., vol. 49, no. 4, pp. 1605–1610, Feb. 2010.

[9] A. N. Manzano Martínez, K. M. P. van Eeten, J. C. Schouten, and J. van der Schaaf, "Micromixing in a Rotor–Stator Spinning Disc Reactor," Ind. Eng. Chem. Res., vol. 56, no. 45, pp. 13454–13460, Nov. 2017.

[10] A. W. Kaandorp, H. Cerfontain, and F. L. J. Sixma, "Aromatic sulphonation V: A kinetic study on the sulphonation of toluene, ethylbenzene and isopropylbenzene in aqueous sulphuric acid," Recl. des Trav. Chim. des Pays-Bas, vol. 82, no. 6, pp. 113–120, 1963.

573296 Intensification of Reaction-Diffusion Processes By Magnetic Nanomixing

Intensification of reaction-diffusion processes by magnetic nanomixing

The ability to control the spatial distribution of concentration in a chemical process is a fundamental requisite for the safe, reliable, economically feasible and environmentally benign operation of most technologies that provide the material basis on which the world depends. The industrial production of everyday materials and commodities such as medicines, cosmetics or electronics would not be possible without effective mixing. The processes that occur in living organisms at both tissue and cell level are also crucially dependent on the local rate of convection and diffusion transport. The influencing of the local diffusion rates or other process parameters that depend on diffusion such as inter-phase mass transfer and reaction kinetics in both chemical and biological systems are also of key interest.

In this study, we are investigating the increase of the local diffusion coefficient by low-frequency magnetic field stimulation of magnetic nanoparticles (magnetic nanomixing). For the creation of a sinusoidal magnetic field, an air-cooled solenoid was fabricated from a broken ferrite ring (gap diameter – 4 mm). We were able to obtain magnetization up to 30 mT at 10,000 Hz. The diffusion rate of a model dye (carboxyfluorescein) in a thin square capillary (2 mm in diameter) was used to evaluate the effect of magnetic nanomixing. The capillary was filled with a solution of dextran-coated magnetic nanoparticles (concentration less than 0.1mg/ml), and 20 ml of carboxyfluorescein was placed in one end. The flow of the aqueous environment due to the concentration gradient or mechanical perturbations was suppressed either by increasing its viscosity (addition of starch) or by immobilization by an agar gel. The capillary was illuminated by a UV-light (390 - 400 nm Dino-Lite microscope) at which the carboxyfluorescein emits green light. The obtained images were evaluated by image analysis (software ImageJ) to get space- and time-dependent concentration profiles of the model dye. The diffusion coefficient was evaluated from this dependency using the Fick's second law. The effect of magnetic particles concentration, magnetic field strength and frequency on the diffusion rate enhancement was experimentally investigated.

579586 Blending in Above Ground Storage Tanks with Side Entering Agitators: Effect of Scale and D/T

The blending of fluids in large, above-ground storage tanks is widely practiced in the petroleum industry yet little research exists of this mixing operation. The most demanding case exists when liquids have stratified into light and heavy layers. The time it takes for the density of the fluid to become axially homogeneous is defined as the blend time. Brine and water have been used as simulants for the light and heavy phase, notably in Wesselingh's work using conductivity to determine the blend times (Wesselingh 1975). Although Wesselingh only explored one propeller geometry, several important geometrical properties were explored, including ratio of propeller vessel diameter and liquid depth to vessel diameter. This study is a continuation of recent studies (Giacomelli et al. 2014) (Grenville et al. 2018) (Boyer et al. 2018) wherein Wesselingh's conductivity technique is applied, but expounds on the geometric impacts including impeller diameter, tank diameter (scale), total liquid height, brine height. Notably, the mixing vessel diameters are varied between 4', 6', 12' to capture scale effect. Blend time predictions are traced to distinct regimes; the regimes are best distinguished using Richardson number, = Δ/ . The implications for scale up and geometric alterations on blend time are discussed.

567338 Characterization and Analysis of Mixing in a Soft-Elastic Reactor (SER)

Characterization and Analysis of Mixing in a Soft-Elastic Reactor (SER)

Jie Xiao*, Minghui Liu, Chao Zou,

Guillaume Delaplace, Romain Jeantet, and Xiao Dong Chen

*Presenting author's e-mail: jie.xiao@suda.edu.cn

International Associated Laboratory - FOODPRINT (INRA Villeneuve d'ascq- Soochow University-Agrocampus Rennes), School of Chemical and Environmental Engineering, College of Chemistry, Chemical Engineering and Materials Science, Soochow University, Suzhou, China, 215123

Abstract

Inspired by animal digestion systems, a new soft-elastic reactor (SER) has been developed in our group[1]. Different from convectional reactors, this type of reactor contains no separate moving parts inside the container and mixing can be achieved through elastic wall movement. SERs potentially have some advantages over conventional rigid wall reactors especially when dealing with highly viscous or corrosive fluids.

In this talk, we will introduce the prototype of this new reactor, specifically focusing on an image analysis based method developed for quantitative evaluation of the mixing performance of SER[2]. The source data collected are RGB values in each pixel of consecutive colour images taken by a CCD camera. The method can determine global indicators such as the mixing time and the average mixing level, i.e., the degree of homogeneity, uniquely characterized by the "distance" away from the ideally mixed state. More importantly, it allows tracking of localized mixing performance throughout the entire mixing process by quantifying spatial and frequency-domain distributions of the mixing level. This simple, generic, and cost-effective method should be a powerful tool for mixing performance evaluation of most reactors with a variety of mixing mechanisms. Furthermore, based on the introduced method, improved designs of SER have been identified, which will also be discussed in this talk.

References:

1. Liu MH, Xiao J, Chen XD. A soft-elastic reactor (SER) inspired by the animal upper digestion tract. *Chemical Engineering and Technology* 2018; 41(5): 1051-1056.
2. Xiao J, Zou C, Liu MH, Zhang GD, Delaplace G, Jeantet R, Chen XD. Mixing in a soft-elastic

reactor (SER) characterized using an RGB based image analysis method. *Chem. Eng. Sci.* 2018; 181: 272-285.

572248 Towards an Understanding of Fluid History in a Turbulent Mixing Tank Using Large Eddy Simulation Model

Many processes such as fermentation, crystallization, and chromatographic resin preparation require sufficient mixing to ensure production, homogeneity, and suspension but are sensitive to high levels of shear due to mechanical agitation. Process scaling can be particularly problematic due to differences in the distribution of shear and bulk mixing during scale up that is hard to replicate at smaller scales. Computational fluid dynamics can be utilized to understand the full 3D process behavior during scale up/down of such processes to debottleneck process development, improve scale up robustness, and develop engineering understanding of fundamental processes.

The typical analysis of strain or energy-dissipation utilizes histograms or percentiles of strain/energy-dissipation in the tank on a spatial basis, i.e. time-averaged. A Reynolds Averaged Navier-Stokes (RANS) model can only provide a time-averaged view, or a pseudo-time-averaged view when using unsteady-RANS, of the process. A large eddy simulation (LES) model provides information about the largest scales of motion and therefore can be utilized to start analyzing the time history of the fluid. This analysis has been difficult previously due to inadequate computational resources to fully sample fluid trajectories in a mixing tank.

We will present the time history of strain/energy-dissipation in mixing tanks across multiple scales using the LES model. First, the time-averaged strain/energy-dissipation predictions will be compared to literature reported values, and the accuracy and limitations of the LES modelling approach will be discussed. The strain/energy-dissipation history can be filtered through signal processing operations that leads to an understanding of power being applied to the fluid at different scales. We will show that there are fundamental relationships between impeller speed and scale that can be used to understand differences in local strain/energy-dissipation.

571617 CFD Modelling and Matlab Analysis to Asses Mixing Performance in Custom Vessel Configurations in the Biopharmaceutical Industry

Title:

CFD Modelling and MatLab Analysis to Assess Mixing Performance in Custom Vessel Configurations in the Biopharmaceutical Industry

Abstract Text:

As the biopharmaceutical industry moves towards single use bag systems which don't have conventional baffling, traditional scale-up equations are becoming less applicable. Large variation is observed in power numbers of agitators at different agitation speeds because of a vortexing impact and the unique design of vessels. The bespoke nature of systems increases difficulty of scaling and many companies have reduced the scale factor from lab to pilot plant to final production scale in an effort to overcome scaling issues.

Numerical modelling, using Computational Fluid Dynamics (CFD) and MatLab was applied to characterise a broad range of biopharmaceutical vessels from upstream bioreactors and wave reactors, to downstream tangential flow filtration vessels and holding vessels. The scales varied from 20 L to 3000 L. An accurate characterisation of the unique vessels allowed the most appropriate scale-up strategy to be identified and tested in CFD, reducing the number of validation batches required, during tech transfer to plant.

557884 Spiroid Design Optimization in a Rotating Continuous Bioreactor for Enhanced Oxygen Mass Transfer

Research on continuous bioreactors requires new methods to improve mixing and mass transfer to increase cell production. The high shear stress from mechanical impellers typically results in greater cell death, especially for mammalian cells. An internal spiroid in a rotating continuous bioreactor provides an alternative for enhanced oxygen transfer and mixing to produce more viable cells. The tubular spiroid collects liquid and gas from the partially-filled rotating bioreactor near the reactor exit and delivers the gas and liquid to the reactor entrance. Computational fluid dynamic simulations were utilized in determining the optimum dimensions of an internal spiroid. The mass transfer coefficients were estimated for the reactor with and without the spiroid using theoretical models to determine the projected increase in oxygen transfer. These values were compared to published experimental values at various rotational rates. Calculated volumetric mass transfer coefficients provided a basis for optimizing the design of the spiroid. This study also determined the dependency of oxygen transfer rate and mixing on the diameter, length and other significant factors of the spiroid at several bioreactor rotational rates. The design of the spiroid included the changing fluid properties due to cell growth.

568506 Effect of Gas Sparge Type on Oxygen Mass Transfer in an Agitated Vessel (comparing CFD versus Experimental Modeling Techniques)

Mixers used in minerals processing often require the suspension of solids in combination with gas dispersion and oxygen mass transfer. Some examples of these types of gas-liquid-solid applications include cyanide destruction, cyanide leach tanks, biological leaching, neutralization, and pre-aeration, among others.

Gas is typically injected in a vessel through a sparge system. As presented previously (Richard Kehn and Antonio Iniguez, "Improvements in Mass Transfer by Sparge System Design Optimization for Minerals Processing," Society of Mining Engineers Conference, Minneapolis, MN, February 2018), it has been shown that an agitated vessel operating under the same air injection rate and gassed power density will yield different oxygen mass transfer rates when different sparging devices are used. The observed change in oxygen mass transfer is due to how the gas interacts with the impeller system, with the physical dispersion and bubble size changing as sparge type is varied. In addition, the mechanical response of the agitator system will also vary by a measurable amount depending on the sparge system chosen. Knowing these variables can allow a plant to optimize both the performance of the agitator system while also considering the cost to implement. The experimental set up consisted of a 1.1 m clear acrylic tank with a flat bottom and four (4) anti-swirl wall baffles and a 406 mm diameter Lightnin A320 wide blade hydrofoil. The wide blade hydrofoil design is typically used for large gassed reactors in minerals processing.

Performing CFD modeling on two phase systems is particularly challenging. It has been shown in previous work (Christopher Tyler and John A. Thomas, "Bubble-Scale Modeling of Gasified Reactors." AIChE Annual Meeting, Pittsburgh, PA, October 2018.) that the Lattice Boltzmann approach can be used to predict both oxygen mass transfer rates and physical dispersion phenomena. Within this approach, the fluid is modeled as a continuous phase and the bubbles are modeled explicitly as Lagrangian point objects that dynamically break-up and coalesce in response to varying local fluid environments. This approach provides direct access to the bubble size distribution and any spatial variations in the gas transfer rates.

In this work, the Lattice Boltzmann modeling technique is used to model three sparge types: an open pipe versus a plate sparge located above the open pipe versus a ring sparge. The CFD modeling results are then compared to the experimental modeling results for validation and comparison purposes.

569929 Multiphase Reacting Flow Simulations and Optimization of Commercial-Scale Aerobic Bioreactors

Aerobic bioreactors are widely used in the synthesis of value-added products in pharmaceutical and biotechnology industries, wherein microbial bioreactions aid in the conversion of sugars to macromolecules. These reactors are also being actively investigated as a cost-effective pathway for the production of low-value commodities such as biofuels [1] and animal feed [2]. Reactor scale-up is one of the important challenges when designing these systems. Gas and liquid phase transport, mass-transfer, and mixing physics at large length scales can significantly affect microbial conversion rates, particularly when the microbial reaction requires a narrow set of conditions. These phenomena are difficult to study in small-scale bench-top reactors that are typically well-mixed. Predictive computational fluid dynamics (CFD) based simulations can therefore aid in the design and optimization of these reactors.

This work presents multiphase Euler-Euler CFD simulations of commercial-scale (~ 500 m^3) aerobic bioreactors. Our mathematical model treats the gas and liquid as interpenetrating phases. This approach reduces the computational complexity of tracking individual gas bubbles that are several orders of magnitude smaller than reactor dimensions. We solve the Reynolds averaged Navier-Stokes (RANS) multiphase equations that account for phase and chemical species transport, interphase mass and momentum transfer and uses a phenomenological model for oxygen uptake by microbes. We use a customized solver derived from open-source CFD toolbox, OpenFOAM [3], to perform these simulations, which has been validated against small-scale reactors in our previous work [4]. This work examines the performance of three different reactor designs, viz. bubble column reactor, airlift reactor with an internal draft tube, and a stirred-tank reactor with Rushton impellers. Reactor oxygen mass-transfer coefficient, gas hold-up, and oxygen distribution are critically analyzed among reactors, and sensitivity studies pertaining to gas flow rates and reactor geometry will be presented.

[1] Humbird, D., Davis, R., and McMillan, J., Aeration costs in stirred-tank and bubble column bioreactors, Biochemical Engineering Journal, 127, 161—166, 2017

[2] Construction complete at Unibio/Protelux commercial scale methane to protein plant," Biofuels Digest, Sept. 2018; "LanzaTech, Shougang open sustainable ethanol refinery in China," Renewables Now, June 2018

[3] Weller, H., Tabor, G., Jasak, H. and Fureby, C., A tensorial approach to computational continuum mechanics using object-oriented techniques, Computers in physics, 12, 6, 620--631, 1998

[4] Rahimi, M., Sitaraman, H., Humbird, D. and Stickel, J., Computational fluid dynamics study of full-scale aerobic bioreactors: Evaluation of gas–liquid mass transfer, oxygen uptake, and dynamic oxygen distribution, Chemical Engineering Research and Design, 139: 283–295.

579291 Modeling Mixing in Aerated Systems: Mass Transport and Flooding

Aerobic and microaerobic fermentation processes require significant aeration for proper organism performance in both stirred tanks and air-lift columns. Aeration provides oxygen for fermentation as well as mixing for the bulk fermenter. Air flows are high and have a significant impact on the overall fluid flow. In the case of air-lift columns, air flow is the only mechanism for fluid flow.

Fermentation productivity is often limited by the rate of oxygen transport into the system, characterized by the bulk mass transfer coefficient $k_L a$, a product of the local mass transfer coefficient k_L and the interfacial area a, both strongly dependent on the turbulent dissipation. Both terms are difficult to predict by standard CFD methodology.

Oxygen transport also depends on the partial pressure of oxygen in the bubbles. In bubble columns, the oxygen content decreases as a bubble rises through the column and oxygen is transferred into the liquid phase, and the pressure decreases as well. Both effects contribute to the partial pressure of oxygen and need to be considered in predicting oxygen transport.

In this work, we use lattice-Boltzmann simulations to model gasified reactors with a two-way coupling between the fluid and gas bubbles. Within the context of this solver, it is practical to track the trajectories of hundreds of millions of individual bubbles—populations that are comparable the total number of bubbles within typical industrial-scale gasified systems. We compare gas hold-up and free surface height in response to gas injection for airlift columns. For agitated systems, we compare predicted power draw, residence time distribution, and interfacial area to experimental values. We show that we can predict the onset of flooding. We conclude by discussing means of modeling explicitly the oxygen transport, accounting for changes in partial pressure due to hydraulic head and oxygen consumption, and compare the results to experimental data.

573996 Use of the Hierarchical Parcel Swapping Method to Simulate Turbulent Subgrid Reacting Flows with Variable Schmidt Numbers

Turbulent reacting flows are ubiquitous in Chemical Engineering and a great deal of work has been performed to understand and quantify them. This is an extremely challenging problem, however, due to the complexity of turbulent flows. Such flows exhibit a wide range of length and time scales. Resolving all continuum scales is not possible for engineering applications due to large Reynolds numbers. The computational cost of direct simulations that do resolve all scales increases with the cube of the Reynolds number. Large Eddy Simulations (LES) overcome this restriction by only resolving the large scale eddies, while modeling the subgrid scale mixing and reaction processes. The key challenge for LES is developing accurate subgrid mixing models, and there are a number of models of varying success that have been published. These include the Interaction by Exchange with the Mean (EIM) model, the Euclidian Minimum Spanning Trees (EMST) model, Curl's model, and the Shadow Position Mixing Model. The central challenge of these models is representing the mixing processes in a manner that is computationally efficient while retaining as much physics as possible. We present a new mixing model termed Hierarchical Parcel Swapping (HiPS). HiPS consists of a collection of notional fluid parcels with adjacency defined using a binary tree. The levels of the tree correspond to geometrically-spaced length scales, the number of which are user-defined. The mixing and pairing and parcels is done at rates consistent with inertial range turbulent scaling, and this is the key physical processes enabling the potential improvement of HiPS over other models. The model has aspects in common with the Linear Eddy and One Dimensional Turbulence Models, but more efficient and flexible, and does not directly imply diffusive mixing on a physical domain. We present an overview of the HiPS model and present results of application to reacting flows. A new formulation of the model allowing species with widely varying Schmidt (Sc) numbers is presented. Variable Sc number scalar transport is important in many applications including combustion, and aerosol formation and transport, including soot formation. Results are presented for multiple scalar species with Sc greater and less than unity, including results from scalar energy spectra. Opportunities for use as an efficient and accurate sub grid scale model are presented, including sub grid flow structures such as jets or near wall closures.

574381 Computational Fluid Dynamics Simulations of Mixing in the Transition Region

Computational Fluid Dynamics Simulations of Mixing in the Transition Region

Zachary J. Oliver, Riley D. Flower, William T. Funkenbusch, Paul W. Steve and David G. Foster

Department of Chemical Engineering

University of Rochester

Rochester, New York

The simulation of mixing in the transition region has both industrial importance and intellectual challenges. Many processes that are operated in the transition region are by default simulated in the laminar or turbulent regions. This paper will overview our work using Computational Fluid Dynamics (CFD) to understand the importance of mixing in the transition region in with regard to several geometries and configurations with fluids of industrial importance. This work will summarize our research to understand CFD simulations in the transition region including the effect of impeller geometry, baffle placement, fluid properties and velocities.

568846 Using GPUs to Run Large Eddy Simulations of Blending in Oil Storage Tanks

In large oil storage tanks, blending is typically accomplished using side-entry mixers positioned inside manholes near the bottom of the tank. These impellers have diameters that are very small when compared to the overall tank geometry and have tip speeds that are high compared to the tank-average flow velocity. This net effect of this disparity is a spatially varying flow field that is turbulent near the impeller but may be laminar in the far-field. From an analysis perspective, these complexities make a priori blend-time predictions difficult and necessitate scale model testing of individual confirmations. From a CFD perspective, transient modeling approaches must be applied to properly predict the spatially varying blending characteristics of such systems.

In this work, we show how GPUs (as opposed to CPUs) can make large eddy simulations of such systems practical and timely. We begin by introducing the concepts governing GPU-performance, as applied lattice-Boltzmann and LES simulation. We then show how, given identical CFD algorithms, a single scientific GPU can execute simulations 100x faster than a single CPU. The extension to multiple GPUs is also discussed and demonstrated. We then use this approach to simulate the start-up and steady state flow behavior of an oil storage tank agitated by a small side-entry mixer. We model both single and stratified two-fluid miscible systems, and compare predicted blend time predictions to experimental measurements.

576402 The Effect of Grid Cell Type in Mixing Time Simulations in Non-Turbulent Stirred Vessels

A new series of CFD simulations were carried out to study mixing times in stirred vessels operated in the laminar and transitional regimes. The focus of the simulations was on the effect of using various types (tetrahedral, hexahedral and polyhedral) and numbers of grid cells on accuracy and numerical diffusion. Results are presented with respect to flow fields and mixing times in vessels stirred by either a centrally positioned Rushton turbine or by three pitched blade turbines on a single axis. The simulations results for the Rushton will be assessed by means of the mixing time data reported by Hoogendoorn & Den Hartog (1964, 1967). In addition, in the simulations for the three pitched blade turbines, the single axis with the turbines was placed at different degrees of eccentricity to study its effect on mixing times.

569149 Analyzing Reactive Mixing Processes Using Computational Fluid Dynamics and *Z*-Transform

Computational fluid dynamics (CFD) is used in the chemical industry to analyze the performance of mixing and reactive processes. CFD simulations complement experimental studies by making accessible measurements that are physically difficult to obtain, and they can also be used in lieu of experiments to conduct investigations where there are safety concerns or equipment and capital limitations. The long computational time required to perform good quality CFD simulations of complex systems is often a necessary cost. Motivated by the desire to reduce computational time and enhance the modeling toolset, we have in recent years demonstrated the use of CFD with *z*-transform, the discrete-time counterpart of the Laplace transform, to provide a powerful, flexible, and time-saving approach for simulating and analyzing the dynamics of mixing processes. Examples that we have illustrated include the open-loop response of a lumped-parameter mixing process, the open-loop response of a distributed-parameter mixing process, the closed-loop response of an unstable mixing process, and the general deconstruction and characterization of mixing processes [AIChE Annual Meeting Papers 2014-354913, 2015-404589, 2016-449915, 2017-486416, and 2018-538613].

In our latest work, we apply the *z*-transform technique to CFD simulations of reactive mixing unit operations. By postprocessing the CFD simulation data in *z*-space, we can construct transfer functions for the transported reactive chemical species that are, by design, functionally equivalent to the linear finite-volume discretized equations that comprise the full-scale CFD model. These transfer functions can then be used to generate additional data for analysis with much less computational effort. In this presentation we illustrate this technique using an example of a continuous stirred-tank reactor (CSTR) with a first-order chemical reaction and show that the results obtained using faster *z*-transform technique are practically identical to the results obtained using fully detailed CFD simulations.

572345 Modeling Polymer Fluid Flow and Reaction with Computational Fluid Dynamics (CFD)

Our research in the CFD simulation and modeling of complex polymeric fluids started in the area of turbulent drag reduction, a phenomena known for over 7 decades. A small quantity of flexible polymer additives can drastically reduce the friction loss of turbulent flow. Drag-reducing polymers are widely used in the transportation of crude oil as well as in the drilling and oil recovery of reservoirs. We use direct numerical simulation to study the fluid flow of such dilute polymer solutions with a viscoelastic constitutive equation, from which much fundamental insight into the polymer effects on turbulent structure and dynamics are obtained. Our recent efforts have expanded to the CFD modeling of polymer reaction engineering, where the complex coupling between mixing, heat transfer, and reaction kinetics is critical to the control of process outcome. Applying CFD to a stirred-tank reactor, we are studying the effects of process design and operating conditions on the conversion rate, polymer molecular weight distribution, and monomer composition in copolymers. This presentation will give an overview of our research efforts in these areas.

555711 CFD-Based Multi-Objective Optimization of Dual-Impeller Configurations in a Gas-Liquid Stirred Tank

CFD-Based Multi-objective Optimization of Dual-Impeller Configurations in a Gas-Liquid Stirred Tank

Miao-Na Chen, Jia-jun Wang*, Xue-ping Gu, Lian-fang Feng

State Key Laboratory of Chemical Engineering, College of Chemical & Biological Engineering, Zhejiang University, Hangzhou P.R. China 310027

jiajunwang@zju.edu.cn

Abstract Gas-liquid stirred reactors are widely used in the process industries owing to the remarkable advantages, such as strong operation flexibility, excellent mass transfer effect and high mixing efficiency. The configuration optimization of aerated stirred tanks is of great significance since configurations are important factors that affect the flow and mixing in chemical reactors, further influencing the mass transfer and reaction process.

The research on configuration optimization of gas-liquid stirred tanks has developed from early experimental measurement to current computational fluid dynamics (CFD) simulation. Although predictions can be made under any operation conditions with CFD method, the process requires trial and error, leading to heavy workload and time cost. Besides, a large amount of experiments will be involved to take into account multiple design variables, which can generally result in local optimum. How to improve the computing efficiency and obtain the global optimum designs still remain very challenging. However, the fast non-dominated sorting genetic algorithm (NSGA-II) offers an attractive alternative to analyze multi-objective problems in a cost-effective way and find global optimum solutions with parallel-searching and fast convergence.

In our work, a multi-objective optimization methodology based on CFD and NSGA-II approaches was developed for dual-impeller design in a gas-liquid stirred tank. The strategy enabled CFD to make predictions for flow field information. It also utilized NSGA-II to cut down the computational demand and acquire global optimum solutions. A bubble size model was introduced to describe bubble size variation in CFD simulation. Six geometrical parameters of impeller blades were selected as design variables. In view of the mass transfer efficiency and energy saving, the overall interfacial area and power consumption were chosen to evaluate the impeller performance.

Model validation was successfully conducted by comparing the experimental results with the simulated data. A Pareto front was reached after 600 iterations to determine the final dual-impeller designs. According to the optimal designs selected from Pareto front, it was found that the pitched concave blade disk turbine (PCBDT) as the lower impeller and down-pumping pitched blade turbine (PTD) or up-pumping pitched blade turbine (PTU) as the upper impeller achieved the relatively larger gas-liquid contact area with much lower energy consumption.

The influence of impeller combination on hydrodynamic characteristics was also investigated through the comparison of nine unique combinations. The results showed the PTD combinations exhibited the best gas dispersion performance, while the PTU combinations showed the worst gas distribution uniformity. The bubble size distribution of each combination shared the same trend: the bubble size varied along the direction of discharge flow, first smaller bubbles were generated in the impeller region and then the bubbles became larger away from the impeller region. And the bubbles near the free surface and the center of fluid circulation loops were relatively large. Moreover, the optimal PCBDT-PTU (lower impeller-upper impeller) presented the highest peak of local interfacial area while the optimal PCBDT-PTD obtained the most uniform interfacial area distribution. Both optimal designs were energy-efficient in pursuing the best overall mass transfer performance.

Experimental measurements of oxygen transfer coefficient and power consumption by oxygen electrode and torque sensor were utilized to evaluate the optimal design. Compared with the reference design with the largest oxygen transfer coefficient, the optimal PCBDT-PTU has improved the oxygen transfer coefficient by at least 11% with a 25% reduction of power input; while the power consumption of the optimal PCBDT-PTD has been decreased by 27%.

The successful establishment of the multi-objective optimization strategy for the impeller design indicates a new application for CFD-based optimization methodology. The automatic optimization stratery show great potential for many industrial fields involving gas-liquid stirred reactors, like fermentation, PX oxidation, and hydrogenation.

562059 Assessment of 3D-PTV to Measure Lagrangian Flow Fields in Stirred Tanks

Background and objectives

3D Particle Tracking Velocimetry (3D-PTV) is an alternative technique to Particle Image Velocimetry (PIV) and Laser Doppler Anemometry (LDA) for visualization of flow fields. It is gaining increasing attention in industry and academia because it enables to measure all the three dimensions of the velocity field and it can obviate the need for laser illumination. 3D-PTV has been carried out using multiple or single camera setups for many applications, including turbulence measurements (Luthi et al., 2005; Liberzon et al., 2012), pulsed impinging jets (Hwang et al., 2007), pipe flows (Oliveira et al., 2015), simulated flows (Pereira et al., 2006), flows in a lid-driven cavity (Kreizer et al., 2010), aortic flows (Gulan et al., 2012; Gallo et al., 2014), plane Couette flows (Krug et al., 2014) and industrial burners (Hardalupas et al., 2000). However, today, there is still a deep lack of information in the literature on the use of 3D-PTV to fluid mixing in stirred tanks (Alberini et al., 2017), particularly in the transitional regime.

3D-PTV involves the addition of tracer particles (in the order of 10 – 1000 μm) within the fluid under investigation, contained in a transparent geometry. A stereoscopic image acquisition system is used to capture synchronous image sequences of the particles in different image planes. After a calibration routine, the 3D coordinates of the particles are computed. Particles are individually tracked from frame to frame, and a set of Lagrangian trajectories in the 3D space is obtained. Eulerian information can also be obtained via post-processing of the time-resolved three-dimensional velocity vectors (Alberini et al., 2017). This requires reconstructing as many trajectories for as long as possible. However, unavoidable unsolved ambiguities result in a certain probability of interruption of the trajectory at each time step (Willneff and Wenisch, 2011). An exhaustive discussion of the principles of PTV may be found in Maas et al. (1993) and Malik et al. (1993).

This work consists of two experimental parts, conducted with a single camera 3D-PTV setup (see Alberini et al., 2017):

- experiments with artificial trajectories with velocities up to 1.9 m s^{-1}, to assess accuracy, precision, repeatability and effect of procedural variables;
- experiments in a 7 L square tank operated in a range of Re from 6,000 to 36,000, to investigate the effects of particle concentration and frames per second on the distribution of the trajectories time scales. To authors' knowledge, this is the first time that a similar analysis is conducted to investigate the tracking performance of 3D-PTV for agitated vessels.

3D-PTV experiments with artificial trajectories

A rotating geometry has been designed to produce artificial trajectories. It consists of four cylinders of different diameter with four control dots. When it rotates at a speed , the uniform circular motion of the dots is known, and the 3D-PTV measurements of the dot velocities can be compared with their expected values ().

The experiments were conducted at 120, 200 and 360 rpm, producing a maximum velocity of 1.9 m s^{-1}. Two half revolutions of the geometry were recorded at 3600 fps. Four circular trajectories, one for each control dot, were obtained. The trajectories extended within the angle where the control dots were visible.

The peaks of the measured velocity were individually fitted with a Gaussian curve, defined by the mean, μ, and the standard deviation, σ. The accuracy of the measurements was assessed considering the relative error between μ and , while the precision was measured by σ. The results showed that accuracy and precision are reasonable in the range of velocities investigated.

The repeatability was assessed considering the standard deviation of the mean measurements over three repetitions of the experiment at 360 rpm. The variability of the measurements was satisfactory.

The circular trajectories were divided in three portions. The peaks relative to each portion were fitted individually to assess the effect of the direction of the motion on accuracy and precision. No relevant differences were found.

Two further experiments have been conducted at 360 rpm and the calibration target was placed in two different positions at a distance of 100 mm. The results were then compared in terms of μ and σ. The variability due to the two calibration routines were not relevant.

3D-PTV experiments in a 7 L agitated tank

Particle trajectories obtained from 3D-PTV have been used to reconstruct the flow field of water in a 7 L square vessel (178 × 178 × 220 mm) equipped with a 60 mm Rushton turbine and operated in a range of Re between 6,000 and 36,000. The effects of the particle concentration and the number of frames per second (fps) on (a) the distribution of the trajectory time scales and on (b) the statistic reliability of time-phase averaged flow fields obtained via post-processing of the Lagrangian trajectories have been investigated. This work is aimed at the development of an appropriate operational window within which the PTV can be optimized for future measurements of transitional flows in agitated vessels.

The flow was visualized using neutrally buoyant microspheres (750 – 820 μm). The experimental matrix includes a nine points partial composite central design and two external points. The particle concentration was varied between 0.02 and 0,54 g L^{-1}, the camera frequency between 900 and 3600 fps, and the impeller speed between 100 and 600 rpm, producing a tip speed between 0.3 and 2.5 m s^{-1}. Given the limitations of the camera memory, each experiment was repeated four times to obtain a cumulative time of recording of 4 seconds. The resulting total number of frames went from 3600 to 14400.

At date of writing, the processing of these experiments is ongoing. The outputs will be the three components of velocity in function of the three spatial coordinates and the variable time, grouped in continuous trajectories.

The analysis of the tracking performance at the different operation conditions will consider:

- the total number of trajectories obtained;
- the distributions of the trajectory time scales, namely the population of trajectories tracked for a certain time τ;
- the normalized distribution of the time scales, where the number of observations is divided by the total number of events.

In light of some previous results, it is expected that, due to unsolved ambiguities and trajectory interruptions, the total number of trajectories increases with particle concentration and fps, and the distributions exponentially decrease with τ.

The statistics of such averaged flow fields will be assessed in some sample regions. In particular, the independence of the ensemble average velocity in those regions with respect to the number of data available will be investigated at the different conditions.

Cited references

Alberini, F., Liu, L., Stitt, E.H., Simmons, M.J.H. (2017), *Chem Eng Sci*, 171: 189 – 203.

Gallo, D., Gulan, U., di Stefano, A., Ponzini, R., Luthi, B., Holzner, M., Morbiducci, U. (2014), *J Biomech*, 47: 3149–3155.

Gulan, U., Luthi, B., Holzner, M., Liberzon, A., Tsinober, A., Kinzelbach, W. (2012), *Exp in Fluids*, 53: 1469–1485.

Hardalupas, Y., Pantelides, K., Whitelaw, J. (2000), *Exp in Fluids*, 29: 220–226.

Hwang, T.G., Doh, D.H., Jo, H.J., Tsubokura, M., Piao, B., Kuroda, S., Kobayashi, T., Tanaka, K., Takei, M. (2007), *Chem Eng J*, 130: 153–164.

Kreizer, M., Ratner, D., Liberzon, A. (2010), *Exp in Fluids*, 48: 105–110.

Krug, D., Holzner, M., Luthi, B., Wolf, M., Tsinober, A., Kinzelbach, W. (2014), *Meas Sci Technol*, 25: 1–13.

Liberzon, A., Luthi, B., Holzner, M., Ott, S., Berg, J., Mann, J. (2012), *Phys D*, 241: 208–215.

Luthi, B., Tsinober, A., Kinzelbach, W. (2005), *J Fluid Mech*, 528: 87–118.

Maas, H.G., Gruen, A., Papantoniou, D.A. (1993), *Exp in Fluids*, 15: 133–146.

Malik, N.A., Dracos, T., Papantoniou, D.A. (1993), *Exp in Fluids*, 15: 279–294.

Oliveira, J.L.G., Van der Geld, C.W.M., Kuerten, J.G.M. (2015), *Int J Multiphase Flow*, 73: 97–107.

Pereira, F., Stuer, H., Graff, E.C., Gharib, M. (2006), *Meas Sci Technol*, 17: 1680–1692.

579524 Characterization of Vortex Formation and Blend Time in Unbaffled Tanks

Vortex formation in unbaffled mixing tanks has not been characterized in previous literature studies. This work has characterized vortex formation as a function of the inverse Froude number and Reynolds number. Additionally blend time studies have been performed using conductivity measurements. The data was collected at liquid level heights ranging from 0.75 to 1.75 of the tank diameter, impeller speeds from 150 RPM to 800 RPM, axial and radial impellers, and tank diameters of 8 inches to 2 feet. The following correlation, based on Grenville was obtained: $5.66 = N_p^{1/3} \times Re \times Fo$ where Fo is the Fourier number.

580150 Suspension of Large and Dense Particles in Stirred Vessels: Further Application and Validation of Grenville-Mak-Brown (GMB) 2015 Approach

Stirred vessels are utilized for various chemical and hydrometallurgical processes involving two or more phases. As we know, most processes in the metallurgical industry involve the processing of solid particles in a liquid phase for the purpose of reaction, separation or transport. The design of these solid-liquid processes in stirred vessels operating in the turbulent flow regime with free-settling slurries requires the prediction of the *Just Suspended* condition N_{JS}, the impeller speed at which all particles are in motion [1]. Recently, Grenville et al. [2] [3] [4] presented experimental data measured at three different scales demonstrating success of an energy balance approach to modeling the suspension process. The particle size and density used in the experiments were less than 500μm and less than 2650 kg/m³ respectively. Literature was sourced to find additional datasets that could be used to improve the validity of the model, including transport of solids in pipelines [5] [6] [7] [8] [9]. New experiments including large and dense particles to sufficiently vary the Archimedes number have been conducted. Generally, the correlation has been validated for large and dense particles (2650-11340 kg/m³, 500 μm $< d_{43} <$ 10,000 μm). The GMB formulation has been generalized to account for local concentration effects due to the formation of a slurry cloud height. The effect of scale and scale up as well as vessel shape will also be discussed. The correlation when applied to slurry transport in pipelines has shown similar success in correlating the critical deposition velocity in pipelines.

References

[1] T. N. Zwietering, "Suspending of Solid Particles in Liquid by Agitators," *Chem. Eng. Sci.*, vol. Vol. 8, pp. 244-253, 1958.

[2] R. K. Grenville, A. T. C. Mak and D. A. R. Brown, "A New and Improved Correlation to Predict "Just Suspension" Speed for Solid-Liquid Mixtures in Stirred Tanks," in *Mixing XXII*, Victoria BC, Canada, 2010.

[3] R. K. Grenville, A. T. C. Mak and D. A. R. Brown, "Suspension of Solid Particles in Vessels Agitated by Axial Flow Impellers," *Chem. Eng. Res. Des.*, pp. 282-291, 2015.

[4] R. K. Grenville, J. J. Giacomelli and D. A. R. Brown, "Suspension of Solid Particles in Vessels Agitated by Rushton Turbine Impellers," *Chemical Engineering Research and Design*, 2016.

[5] G. Baldi, R. Conti and E. Alaria, "Complete Suspension of Particles in Mechanically Agitated Vessels," *Chem. Eng. Sci.*, pp. 21-25, 1978.

[6] J. T. Davies, "Particle Suspension and Mass Transfer Rates in Agitated Vessels," *Chemical Engineering and Processing*, pp. 175-181, 1986.

[7] M. J. Cliff, M. F. Edwards and I. N. Ohiaeri, "The Suspension of Settling Solids in Agitated Vessels," *IChemE Symposium Series*, vol. 64.

[8] S. Ibrahim and A. Nienow, "Particle Suspension in The Turbulent Regime: The Effect of Impeller Type and Impeller/Vessel Configuration," *Trans IChemE,* vol. 74, pp. 679-688, 1996.

[9] H. Voit and A. Mersmann, "General Statement for the Minimum Stirrer Speed During Suspension," *Ger. Chem. Eng.* , vol. 9, pp. 101-106, 1986.

573097 An Improved Kinetic Model for Micromixing Characterization Using the Villermaux-Dushman Method

In the last few decades, Process Intensification (PI) and the development of novel equipment have become key approaches for engineers and scientist to tackle the challenges of the modern era. The increasing effects of climate change require more effective measures to limit emissions and develop more sustainable processes. Furthermore, industries have the necessity of reducing waste generation but yet generate profits. However, for new technology or new equipment to get into full scale implementation in industry, sufficient knowledge needs to be acquired both in potential applications as well as in economic feasibility.

In many industrial processes, reducing energy-intensive separation steps during purification of final products could be achieved by enhancing the selectivity towards the desired products, or by reducing undesired by-products that in many cases are dangerous or harmful. This is particularly difficult for processes that have fast reactions competing for a limiting reagent, i.e., where micromixing plays an important role.

Micromixing is commonly defined as the homogenization of the system at the smallest scale, where reaction occurs. The process has a characteristic time, commonly called micromixing time, and it can define the product distribution of fast competitive-consecutive or competitive-parallel reactions. Typically, test reactions are used to measure qualitatively and quantitatively the micromixing time in chemical reactors. One of the most extensively employed test reaction systems is the Villermaux-Dushman system, developed in the 1990's [1] and since then used to characterize both traditional and novel reactors[2][3].

The experimental method requires a small injection of acid, which depending on the micromixing efficiency can favor one of the two possible reactions: (1) neutralization of a buffer, or (2) iodide-iodate reaction. However, the only way to quantitatively asses the micromixing time is having enough information about the kinetics, and for some years there has been controversy on whether the kinetic model for the iodide-iodate reaction is accurate enough to quantify micromixing times[4].

In order to allow a quantification of the micromixing times with this method, a new independent study was conducted, taking into account the most recent findings on the topic and addressing the issues of the most used kinetic model. Preliminary results indicate that the partial reaction orders were in agreement with published models[5], but the dependency of the kinetic constant on the ionic strength was overestimated in previous studies[6]. The latter was determined using potassium sulfate to vary the ionic strength together with the adoption of sulfuric acid as a proton source. These two factors led to a "buffer-like" situation, slowing the reaction rate and hence the apparent kinetic constant.

In this study perchloric acid, a strong monoprotic acid, was used as a proton source. Furthermore, the inert salt used to adjust ionic strengths was sodium perchlorate, to avoid undesired protonation of a weak base such as potassium sulfate[7]. Fresh solutions of potassium iodide and potassium iodate were prepared for all the experiments, varying concentrations to study the partial orders and adjusting with sodium perchlorate to adjust to the desired ionic strengths. A small batch reactor with integrated cooling to ensure a constant temperature of 20 °C was loaded with 50 ml of the solution. An injection of 1 ml of perchloric acid at the desired concentration was added to start the reaction while stirring the contents of the reactor with a magnetic stirrer. With in-line UV/Vis spectroscopy the absorption at 353nm was used to determine the concentration of triiodide using Beer-Lambert law.

Certainly the choice of the concentration set represented a challenge during the experimental planning. First, the use of high concentrations leads to a quasi-instantaneous reaction rate which cannot be accurately determined. Second, the concentration of Potassium Iodide has to be sufficiently high to push the equilibrium forward, and convert most of the Iodine into triiodide. It was observed that failing to fulfill this condition caused considerable evaporation of Iodine and a deviation of the observed kinetic constant. Finally, increasing the ionic strength to 1M by adding sodium perchlorate required also an increase of the acid concentration, because the reaction rate was too slow to accurately measure it.

After the above considerations were taken into account, the concentration sets were selected in such a way that reaction rates were fast enough to avoid evaporation of Iodine over time (max. 300 seconds), but slow enough to ensure the observation of a kinetic regime (min. 20 seconds). The observed kinetic constant was fitted using the software Matlab for each experiment. The following correlation between the kinetic constant and the ionic strength was found:

$$r = k_0\,(\gamma_{H^+}[H^+])^2(\gamma_{I^-}[I^-])^2(\gamma_{IO_3^-}[IO_3^-])^1 \qquad (1)$$

$$k_{obs} = k_0\gamma^5 \qquad (2)$$

$$k_0 = 1.007 \times 10^9,\ M^{-4}s^{-1} \qquad (3)$$

$$\log_{10}(\gamma) = -0.3077\left(\frac{\sqrt{I}}{1+\sqrt{I}} - 0.086I\right) \qquad (4)$$

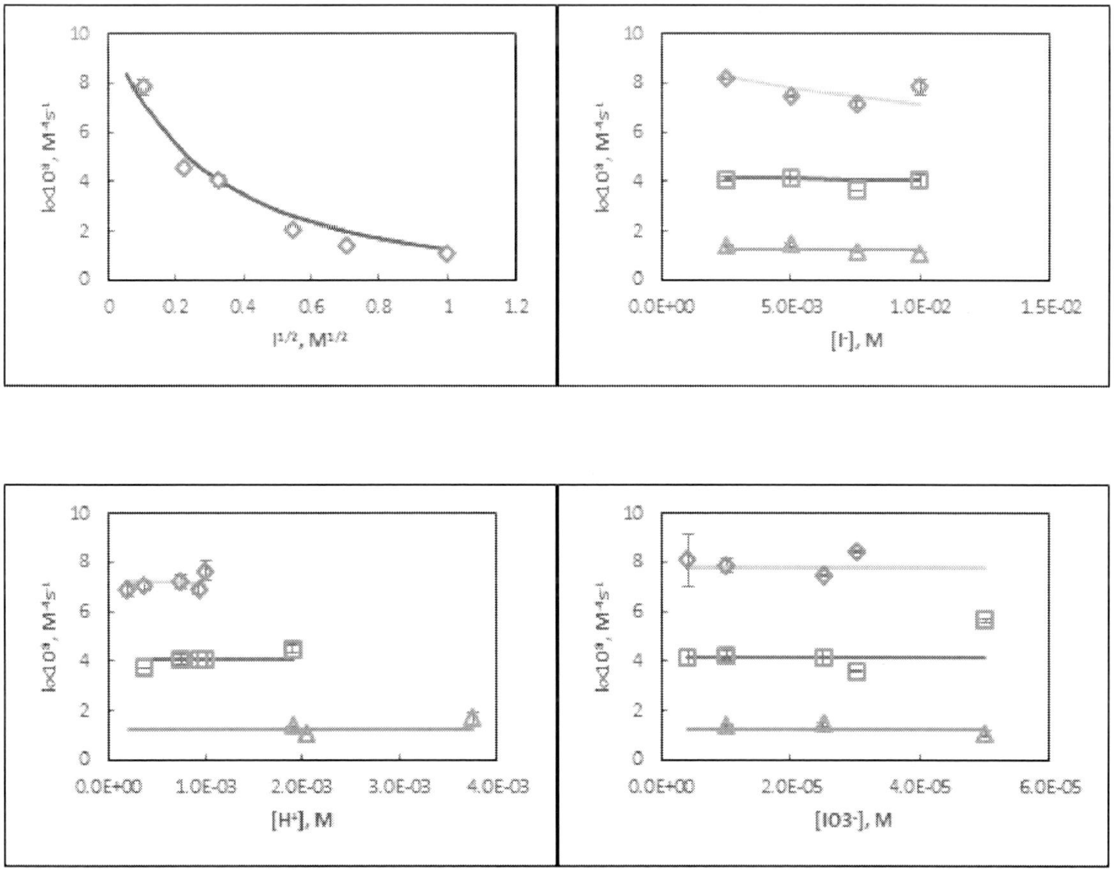

The proposed empirical model (Eq 1) consists on an "intrinsic" value at infinite dilution that decreases as the ionic strength increases due to its impact on the activity coefficients of all the species. The dependency of the activity coefficient on the ionic strength is expressed in the form of a Debye-Hückel equation with one extra parameter, similar to what Davies[8] proposed. This model is in very good agreement, not only with previously published kinetic data [5][8], but also with the most recent studies[10].

With this model, a more accurate estimation of the micromixing times can be achieved, improving the quantitative characterization of novel equipment such as rotating packed beds, spinning disc reactors, rotor-stator mixers, micro-reactors, and more.

Furthermore, a new comparison of micromixing times obtained with another test reaction (the diazocoupling of Naphthols) will be presented, since previously there was a significant disagreement between the two methods[11]. Preliminary results of ongoing research show discrepancies that could be attributed to the different characteristic reaction times more than to the reliability of the systems.

References:

[1] M.-C. Fournier, L. Falk, and J. Villermaux, "A new parallel competing reaction system for assessing micromixing efficiency—Determination of micromixing time by a simple mixing model," *Chem. Eng. Sci.*, vol. 51, no. 23, pp. 5187–5192, Dec. 1996.

[2] M. Assirelli, W. Bujalski, A. Eaglesham, and A. W. Nienow, "Study Of Micromixing in a Stirred Tank Using a Rushton Turbine: Comparison of Feed Positions and Other Mixing Devices," *Chem. Eng. Res. Des.*, vol. 80, no. 8, pp. 855–863, Nov. 2002.

[3] A. N. Manzano Martínez, K. M. P. Van Eeten, J. C. Schouten, and J. Van Der Schaaf, "Micromixing in a Rotor-Stator Spinning Disc Reactor," *Ind. Eng. Chem. Res.*, vol. 56, no. 45, pp. 13454–13460, Nov. 2017.

[4] J. R. Bourne, "Comments on the iodide/iodate method for characterising micromixing," *Chem. Eng. J.*, vol. 140, no. 1–3, pp. 638–641, Jul. 2008.

[5] D. A. Palmer and L. J. Lyons, "Kinetics of iodine hydrolysis in unbuffered solutions." 1989.

[6] P. Guichardon, L. Falk, and J. Villermaux, "Characterisation of micromixing efficiency by the iodide-iodate reaction system. Part II: Kinetic study," *Chem. Eng. Sci.*, vol. 55, no. 19, pp. 4245–4253, Oct. 2000.

[7] A. Kölbl, M. Kraut, and R. Dittmeyer, "Kinetic investigation of the Dushman reaction at concentrations relevant to mixing studies in microstructured cyclone type mixers," *Chem. Eng. Sci.*, vol. 101, pp. 454–460, Sep. 2013.

[8] C. W. Davies, "Extent of dissociation of salts in water. Part VIII," *Trans. Faraday Soc.*, vol. 28, pp. 2093–2098, 1938.

[9] S. M. Schildcrout and F. A. Fortunato, "Spectrophotometric study of the rate of the aqueous iodate-iodide reaction," *J. Phys. Chem.*, vol. 79, no. 1, pp. 31–34, Jan. 1975.

[10] G. Schmitz, G. Bourceanu, and I. Ungureanu, "Effects of Ce(III) and Mn(II) on the Dushman reaction and simulations of the Briggs–Rausher reaction," *React. Kinet. Mech. Catal.*, vol. 123, no. 1, pp. 81–92, Feb. 2018.

[11] P. Guichardon, L. Falk, and M. Andrieu, "Experimental Comparison of the Iodide-Iodate and the Diazo Coupling Micromixing Test Reactions in Stirred Reactors," *Chem. Eng. Res. Des.*, vol. 79, no. 8, pp. 906–914, Nov. 2001.

574056 Comparison of CFD Simulations to Experimental Results for Blend Time of Turbulent Newtonian Fluids in Stirred Tanks

Blending, the process of combining miscible components into homogeneous solutions, is a common unit operation performed by rotating agitators. Industrialists are interested in minimizing the blend time to decrease operating periods and increase throughput while also minimizing capital and operating costs. Therefore, agitator manufacturers are driven to provide efficient and cost effective solutions for blending applications as well as accurate blend time predictions.

A blend time investigation for stirred tanks was conducted to study the effect of impeller diameter ($D = T/2$, $T/3$, and $T/5$), impeller off-bottom distance ($C = T/2$, $T/3$, and $T/5$), and impeller type (A310 Hydrofoil, Pitched-Blade Turbine, and Rushton Turbine) while considering tank diameter ($T = 0.80$ and 1.22 m) and mean specific energy dissipation ($\varepsilon = 0.005$ and 0.010 W/kg) for fully turbulent Newtonian fluids. The response of conductivity probes to the injection of a saturated NaCl tracer was monitored to experimentally measure blend time for 108 unique configurations produced by a full factorial experimental design. The results of the experimental blend time measurements were analyzed to determine trends and correlation coefficients were fit from the data. A growing body of findings has been presented at the 2017 AIChE Annual Meeting, published in the submitting author's master's thesis, "Investigation of Blend Time for Turbulent Newtonian Fluids in Stirred Tanks", and presented at NAMF Mixing XXVI.

The latest findings presented at NAMF Mixing XXVI suggested that the optimum impeller type for turbulent blending was a function of the impeller diameter to tank diameter ratio (D/T) with an inflection point occurring at a D/T of 0.30. Below 0.30 D/T, the A310 hydrofoil was the most efficient impeller type. Above 0.30, the Rushton Turbine was the most efficient impeller type. At 0.30, all three impeller types were equivalent. This conclusion contradicted the two main turbulent blending theories: the first being impeller type independence and the second being that flow efficient impeller types, like the A310 hydrofoil, are more energy efficient at blending. A mechanism for explaining the findings was not proposed, but a combination of exploratory theoretical derivations and transient CFD simulations with LES filter were proposed to probe for an explanation.

Up to this point, no simulation-based approaches have been utilized in the investigation. Therefore, the first goal is to validate a simulation methodology that may then be used for mechanism exploration. A subset of the experimental plan will be simulated. All three impeller types (A310 Hydrofoil, Pitched-Blade Turbine, and Rushton Turbine) at all three impeller diameters ($D = T/2$, $T/3$, and $T/5$) will be simulated, but impeller off-bottom will be fixed at $T/3$, tank diameter fixed at 1.22 m, and mean specific energy dissipation fixed at 0.010 W/kg. Probe and tracer injection locations in the simulations will be identical to the experiments, and blend time will be determined using identical methodology. The validation criteria will be that simulation-derived blend times must fall within ±14% of the experimentally determined blend times as that was the average relative variation among experimental trials.

The simulation methodology that will be studied is a transient Lattice Boltzmann solver with LES filter. This methodology will be executed using M-Star CFD. The suitability of using M-Star CFD to predict blend time of turbulent Newtonian fluids in stirred tanks will be discussed and next steps will be proposed. Since this modeling approach is physics based, preliminary insights into the flow mechanisms driving this impeller-dependent blending behavior will be presented.

574913 Comparison of Blend Time Measurements: An Experimental and Computational Study

The understanding and prediction of blend times in agitated vessels is industrially important for many applications (e.g. chemical reactions, thermal homogeneity, additive blending). The measurement of blend time can be done in different ways, depending on the application. Understanding the measurement techniques and being able to interpret the information from them is important. Depending on the application, one or more of these techniques can be used to successfully measure the blend time.

Most work in estimating blend times is done with "clean" fluids that minimize noise in the measurement signal. Techniques for measuring blend times include decolourization via chemical reaction (e.g. acid/base with an indicator), colourization (e.g. adding a dye), and temperature or concentration monitoring via probes at particular points in a vessel. Comparing the values of these different techniques can be challenging because of the stochastic nature of turbulent blending and various definitions of "blended". Examining different measurement techniques experimentally simultaneously is time consuming and challenging from an experimental standpoint. With recent increase in computing power, high fidelity CFD simulations that were strongly in the academic domain a decade ago are now available as engineering tools. Using these types of tools to examine blending can help understand the relationships between various types of blending measurements.

The work to be presented will include both experimental and computational elements. Data from experimentation and published literature will be compared with CFD results to form a basis for validation of the CFD. Once validation, the CFD can be used to examine the several different blending criteria simultaneously. This includes acid/base reactions with an indicator, REAX by FMP, and probe-based responses. Because the simulation allows the concentration to be known at all locations within the vessel (with limits at the simulation resolution), comparisons to the uniformity of the entire "population" of the vessel positions will also be done

571969 Controlled Droplet Size for Oil-in-Water Emulsions: Experimental and Modeling Approach to Characterize Work History Applied By Mixing Devices

To reduce the time and cost associated with scaling-up processes for emulsification, we need to fundamentally understand how the different mixers create dispersions and how to control the final dispersion properties. This knowledge will enable flexibility in transferring products by improving emulsion technology selection. The primary goal of this project is to develop a better understanding of drop breakup within the high shear mixers used for emulsification by examining the fundamental relationships between shear input and break-up mechanism and the final drop size/distribution, developing correlations to describe these relationships, and characterizing different high shear mixers using the correlation. M-Star CFD was used to evaluate several mixers and examine the flow behavior of particles as they are mixed. The power dissipation the particle experiences along its path is then used to approximate a maximum drop size and size distribution.

558076 Defining a Scale-up Strategy for an Oil in Water Cosmetic Emulsion

Corresponding author: aorjuelal@unal.edu.co

A major driver in the cosmetic market is the rate at which new and innovative products are designed and engineered to fulfill consumer's needs. As a result, the time required to introduce a product in the market is a paramount objective during product design. Despite different tools have been developed for in-silico product conception and design, and novel technologies enable high throughput prototyping and manufacturing, a major limiting factor remains in the scale-up process. During development of structured products in the cosmetics industry (e. g. emulsions), scale-up to industrial production is a time-consuming, resources-intensive, and complex process, reason why it frequently involves trial and error techniques.

Several variables need to be considered to ensure the product desired properties and its long-term stability. Variables such as mixing times, stirring rates, emulsification temperature, heating or cooling rates, equipment geometry, and ingredients addition procedures play a major impact in the final characteristics of the emulsified products. In the scale-up process all these variables require to be tuned and adjusted, through trial and error methods. This makes this process highly inefficient.

In this direction, different systematic scale-up techniques have been developed for emulsions production. Specifically, for water in oil, the use of process invariants such as some dimensionless numbers of the form ND^a, have been incorporated as key parameters during change of scale. While these techniques mainly focus in maintaining a defined drop size distribution, it has been verified that this doesn't guarantee to maintain all the emulsion characteristics. Additionally, most emulsified cosmetic products are oil in water emulsions, and there is lack of scale-up data in the open literature for this type of products.

Thus, the aim of this work is to study the influence of some process variables (emulsification temperature, stirring rates, cooling rate, and addition time) in the characteristics of an oil in water emulsion (drop size distribution, stability and rheological behavior). These were studied at different processing scales in order to define suitable scale-up strategies.

Experiments were carried out using a rotor stator mixer with tip speed between 0.5 m/s and 2 m/s, an anchor mixer in rates among 0.01 m/s and 0.5 m/s, addition times between 2 min and 9 min and emulsification temperatures between 70°C and 90°C. Additionally, the cooling rate was varied using cooling fluid temperatures between 6°C and 24°C. Based upon the obtained results at the different levels, a scale-up methodology is proposed for the production of oil in water emulsion products, taking into account resources and the time-to-market minimization.

567573 Effective Scale-up of Aerated Fermentation Processes with Complex Rheology to Industrial Scale

In the biopharmaceutical industry, process transfers from pilot scale into production scale are often carried out with large scaling steps, particularly when fast and efficient development cycles are desired and when targeting world-scale plants. An accurate scaling concept is therefore of utmost interest, as inaccuracies may lead to large changes in the process conditions.

One prominent example of a widely used, yet inadequate, scaling concept for aerated bioreactors is based on a constant specific power input and aeration rate. For large scaling factors, this concept leads to an undesired change of the impeller flow regime, e. g., flooding of the bottom impeller. The present contribution introduces a novel scaling approach derived from the full dimensional analysis and mechanistic equations. The concept is based on the adequate scaling of the impeller flow regime and enables scaling the rheology of non-Newtonian fluids through extending the concept with one additional dimensionless number.

A large experimental data set supports deriving reliable correlation equations for the Newton number, gas hold-up and volumetric oxygen mass transfer coefficient (k_La). The data covers a large range of process conditions typical for industrial-scale fermentations, which were scaled down to a Plexiglas 160-L pilot scale reactor equipped with four Rushton impellers. Multiphase GPU-based CFD simulations, as well as real measurements of the flow regime, gas hold-up, and power input at the individual impeller stages extend the new insights into the large-scale liquid and gas bubble flow. The study further verifies the concept of applying a separate data set from the pilot reactor and from two industrial reactors at the 110-m^3 and 170-m^3 scale. The resulting prediction quality is unparalleled, as illustrated through the critical analysis of existing correlations. Thus, the new concept provides adequate correlations for scaling the flow regime and rheology between pilot and industrial scale reactors, verified for a scaling factor of three orders of magnitude.

573760 Direct Numerical Simulation of Bioreactors Using GPUs

Turbulent fluid flow in bioreactors occurs over a range of length and time scales. In the upper limit, these systems contain macro eddies with length scales comparable to the vessel diameter and time scales comparable to the tank blend time. In the lower limit, these systems contain micro eddies with length and time scales characterized by the Taylor and Kolmogorov scales. In chemical reactors and combustion systems, process outcomes are governed by fluid dynamics occurring across the entire spectrum of eddy scales: the macro scales describe how systems blend, while the microscales describe how species will react. From a modeling perspective, direct numerical simulations are required to obtain the complete insights required to properly predict reactor performance.

In this work, we show how GPUs (as opposed to CPUs) can make DNS of industrial mixing systems practical and timely. We begin by introducing the concepts governing GPU-performance, as applied lattice-Boltzmann simulations. We then present various criteria for monitoring the convergence from large-eddy simulation (LES) to direct numerical simulation, within the context of the Taylor and Kolmogorov scales. We then apply these criteria to study blending, energy dissipation, and reaction rates in a benchtop bioreactor in the LES and DNS regimes. The effects of simulation resolution on runtime and results are discussed.

568568 Use of Acoustic Emission for Identifying Gas-Liquid Mixing Regime in Agitated Vessels at Different Scales Applying Machine Learning

Use of acoustic emission for identifying gas-liquid mixing regime in agitated vessels at different scales applying machine learning

Giuseppe Forte, Federico Alberini, Mark J. H. Simmons, E. Hugh Stitt

Johnson Matthey Technology Centre, Billingham, TS23 4LB, UK
School of Chemical Engineering, University of Birmingham, Edgbaston, B15 2TT, UK

INTRODUCTION

Operations involving Gas-Liquid agitated vessels are common in the biochemical and chemical industry; ensuring good contact between the two phases is essential to process performance. In this work, Acoustic Emission (AE), using a piezoelectric sensor, was applied to evaluate gas-liquid mixing regime within two-phase (gas-liquid) and three-phase (gas-solid-liquid) mixtures in a 3L stirred tank equipped with a Rushton Turbine and a ring sparger. AE has been previously applied in similar applications in pipelines [1] and to characterise bubble shape and size [2-4]. While moving up through the vessel, gas bubbles will collapse, break or coalesce generating sound waves transmitted to the wall and to the acoustic transmitter. The system was operated at different flow regimes (non-gassed condition, loaded and complete dispersion) achieved varying impeller speed and gas flow rate, with the objective to use features in the acoustic spectrum to univocally identify the different conditions. The methodology was then applied to systems at larger scale, 10, 100 and 1000L to assess performance at manufacturing volumes.

MATERIAL AND METHODS

The used apparatus was assembled as for similar works in the literature [5]: a piezoelectric sensor (Vallen Systeme GmbH, Icking, Germany), with resonance frequency of 375 kHz and diameter of 20.3 mm, was attached to the tank with a silicone based vacuum grease. The sensor was connected to a preamplifier (40 dB gain, Vallen Systeme) and an oscilloscope (5243A Pico® Technology Limited) controlled by a laptop was used to record data. Measurements at each condition were taken for 0.2 s with a sampling rate of 750 kHz, ensuring coverage of the resonance frequency of the sensor in the spectrum (Nyquist rule). The sensor was placed on the 3 L Perspex vessel outside wall at a height corresponding to the impeller region.

Machine learning was used to process the acquired data applying logistic regression algorithm. The acquired data in the time domain were preprocessed using Fast Fourier Transform, filtered and selected with respect of their variance, before being fed to the machine learning algorithms.

The dataset was divided into a training set, used to train the machine in recognising features in the spectrum corresponding to each condition, and a testing set, unseen by the machine, used to evaluate accuracy of regime prediction. The training dataset comprised of runs at different regimes in the bi-phasic condition (gas-liquid), while in the test, the system was challenged to recognize the same conditions in the two-phase case and in case of three-phase mixing, where stainless steel particles (3-5 % w/w, density ρ_s = 8000 kg m^{-3}, size between 0.177 and 0.420 mm) were added to the mixture.

RESULTS

In first instance, for the biphasic case, a dataset composed by 100 spectra was fed to the trained algorithm to be tested. The machine, based on conducted training, assigned to each spectrum a predicted class (ungassed, loading or recirculation). In table 1, the results of the test are reported as number of correctly predicted cases and failures for the biphasic case.

Real condition	Correctly predicted	Failed to predict
Ungassed	35	5
Loading	26	4
Recirculation	29	1
Global	90	10

The five spectra are wrongly classified as loading instead that ungassed and therefore are counted as misclassified. Other few spectra are not classified correctly by the algorithm, that however achieves an accuracy of 90% on the carried-out study.

In the second part of the study, stainless-steel particles were added at different concentration and the acoustic signal was acquired at the three regimes. The obtained results are summarized in Table 2.

Real condition	Correctly predicted	Failed to predict
Ungassed	90	10
Loading	66	4
Recirculation	150	10
Global	306	24

In this case the dataset comprised of 330 spectra and the accuracy increases to 92.3%, with an even distribution of errors across the different conditions. The presence of the solids does not affect the classification algorithm that using the training acquired in the bi-phasic condition, it is still able to correctly identify the regime characterising the analysed spectrum. The reason for this can be found in the influence of the solids on different frequencies from the ones characterising the air bubble disruption and impacts.

CONCLUSIONS

In this work, Acoustic Emission was used in combination with a machine learning algorithm, logistic regression, to identify gas-liquid mixing condition in a stirred tank. The machine was trained to recognize ungassed, loading and recirculation condition using a training dataset, consisting in the acoustic spectrum acquired at the different regimes in the biphasic condition (gas-liquid). The system was then challenged to recognise such regimes both in two-phase and three-phase condition (obtained adding stainless steel particles), achieving an accuracy equal to and above 90% respectively. The developed methodology is extended to larger scale in order to evaluate physical and practical reasons that would undermine its application in industrial environments. The conducted study aims to propose AE as a potential diagnostic and condition monitoring technique for fluid mixing applications in combination with the use of machine learning algorithms, for applications at large scale for process monitoring.

REFERENCES

[1] Addali, A., Al-lababidi, S., Yeung, H., Mba, D., & Khan, F. (2010). Acoustic Emission and Gas-Phase Measurements in Two-Phase Flow. *Proc. Institution of Mech. Eng., Part E: Journal of Process Mechanical Engineering*, *224*(4), 281–290.

[2] Leighton, T. G. The acoustic bubble, 1994 (Academic Press, London).

[3] Manasseh, R. Acoustic sizing of bubbles at moderate to high bubbling rates. *In Proc.4th World Conference on Experimental heat transfer, fluid mechanics and thermodynamics*, Bruxelles, Belgium, 1997, pp. 943–947.

[4] Al-Masry, W. A., Ali, E. M., and Aqeel, Y. M. Determination of bubble characteristics in bubble columns using statistical analysis of acoustic sound measurements. *Inst. Chem. Eng., 2005,83* (A10), 1196–

1207.

[5] A. Nordon, R. J. H. Waddell, L. J. Bellamy, A. Gachagan, D. McNab, D. Littlejohn G. Hayward. Monitoring of a heterogeneous reaction by acoustic emission, *The Royal Society of Chemistry: the Analyst*, vol.129, pp.463-467, 2004.

570985 Hydrodynamics Investigation of USP Dissolution Testing Apparatus I (Basket Apparatus) and Effect of the Basket Mesh Size on Hydrodynamic Characteristics

The United States Pharmacopoeia (USP) develops and disseminates compendial quality standards for drug products and other pharmaceutical products, including the dissolution testing methods routinely used in industry and typically accepted by FDA. The USP Dissolution Testing Apparatus I is commonly used as an analytical tool to evaluate the performance of solid drug products and to determine if a drug formulation is within the acceptance standard criteria. In this test the drug oral dosage form, typically a tablet, is placed in a perforated basket immersed in a small volume-vessel and rotating at constant fixed speed. The tablet dissolution process is followed over time by taking samples from the liquid and determining the dissolved drug concentration. Several reports in the literature have indicated that there is considerable variability in the dissolution profiles obtained with this dissolution testing apparatus. One of the most likely reasons for this is the susceptibility of the apparatus to small variation in the system's geometry often introduced through minor deviations of the actual system from perfect symmetry or the use of baskets of different mesh sizes. The objective of this work was to determine the fluid flow in the dissolution apparatus with standard basket mesh size under different operating conditions using Particle Image Velocimetry (PIV) as an experimental tool and to investigate and compare the velocity distribution profiles obtained using different basket mesh sizes. The results indicate that variation in geometric parameters and operating variables within standard criteria appear to significantly affect the hydrodynamics of these systems contributing to the variability of dissolution testing results. The results of this work can provide a better understanding of the hydrodynamic factors affecting the dissolution process in this apparatus and indicate possible approaches to minimize variation in dissolution test results.

575214 Efficiently Characterize Process Scalability in the Micromixing-Mesomixing Space Using the Bourne Protocol

Reaction and crystallization processes often need to balance competing rates in order to achieve the desired results. Mixing can affect certain rate processes, so understanding the role of mixing is a critical part of process development and scaling. Process developers need efficient ways to understand how mixing matters to their processes.

Since it was introduced in 2003, the Bourne Protocol has been used to efficiently understand the effect of impeller speed, feed time, and feed location on semi-batch or gradual-addition processes. Bourne's simple experimental plan creates a framework to understand if micromixing (mixing at the molecular or particle length scale) or mesomixing (mixing at feed plume length scale) or macromixing (mixing at the length scale of the vessel) matters in a process. Applications and enhancements to the Bourne Protocol have been discussed in a variety of venues in recent years (see References).

This paper explores how the Bourne Protocol enables process developers to define regions of acceptable process performance in the micromixing-mesomixing space. There are many potential undesirable process responses that can be affected by mixing: low yield or conversion, poor selectivity, unacceptable purity, unmanageable particle size, poor crystal morphology, inappropriate polymorph, etc. By looking at examples in the literature in a different way, we will see that regions of different process failures can be plotted in the micromixing-mesomixing space, enabling understanding of the acceptable process operating window. Scalability of mesomixing will be discussed from the crystallization literature. Understanding how a process responds to different levels of micromixing and mesomixing can enable more successful scaling, suggest potential process improvements, and speed troubleshooting of industrial processes.

References:

Bourne J.R., "Mixing and the Selectivity of Chemical Reactions", **Org. Process Res. Des. Dev.**; 2003; 7(4) pp. 471-508.

Sarafinas, A. "Test Process Mixing Sensitivities Using the Bourne Protocol," **Scientific Update Webinar**, 13 November 2018.

Sarafinas, A. "Using the Bourne Protocol to test mixing sensitivities in the lab," **DynoChem Guest Webinar**, 8 August 2018.

Sarafinas, A. "A further look at the Bourne Protocol for efficient investigation of mixing sensitivities during process development," **AIChE 2017 Process Development Symposium**, Toronto, Canada, June 2017.

Sarafinas, A. and C.I. Teich in Kresta, S.M., A.W. Etchels III, D.S. Dickey, V.A. Atiemo-Obeng, **Advances in Industrial Mixing: A Companion to the Handbook of Industrial Mixing**, John Wiley and Sons, 2016.

Teich, C.I., A. Sarafinas, P.M. Morton, **AIChE 2010 Annual Meeting**, Salt Lake City, UT, November 2010 (2 papers):

- "Can this process be saved? A search for understanding using the Bourne Protocol and advanced process development tools"
- "Taking the min to the max: a case study in small scale process development using on-line reaction calorimetry and in-situ particle characterization"

582237 Complexities in Simple Mixing

The mixing in many food systems seems quite simple. Many of the relevant processes are single phase Newtonian fluids, designed for a specific mixing time or residence time distribution. Stirred tanks and stirred tanks in series are common vessels for single-phase reactions such as enzymatic modifications. Even these simple cases can show non-intuitive results. We will discuss three such examples: residence time distributions in an unstirred-tank designed to mix feed streams, residence time in a series of CSTRs approximating a plug-flow reactor, and mixing times in a shear-thinning solution.

561004 Steam Stripping of Aroma from Coffee: Multi Scale Modelling of Extraction Behaviour

Aroma is a major factor in consumer satisfaction in drinking coffee, and more than 1000 compounds have been identified n coffee spanning various chemical groups. During instant coffee manufacture, aroma is extracted from fresh roast and ground coffee, to be added back before the final drying stage, preventing evaporative and thermal losses.

One means of aroma extraction can be steam stripping of a coffee bed, the moisture content of which may range from dry to fully saturated. The product aroma profile is largely determined by this process, therefore, there is a need to understand how key aroma compounds extract upon contact with water and steam at different conditions.

A theoretical model of steam stripping a coffee bed has been constructed, consisting of four stages of mass transfer, at the scale both of the individual particle and the whole bed; oil-to-water aroma release; diffusion through the porous coffee grains; partition into the gas phase; and advection up the packed column. Simulations showed three rate-limiting extraction mechanisms related to oil-water partitioning, diffusivity and water-air partitioning.

Experimental data has provided evidence for these three extraction behaviours but there are compounds limited by other factors not anticipated by the model. These are believed to arise from interactions with the soluble coffee components and the solid coffee matrix, a hypothesis supported by further experiments.

Process variables including the grind (particle) size, moisture level and distribution in a coffee bed have been tested and reveal a range of responses depending on the chemistry of the aroma. Moisture may help mobilize the soluble residues on the surface of the coffee grain which may contain aroma and thus assist its extraction, or it may enable the aforementioned interactions to take place, thus limiting volatization of aroma.

Experimental and modelling results have been combined to reveal a wide range of extraction kinetics of volatile aromas in three phase steam stripping of roast and ground coffee. This knowledge reveals how manipulation of the process conditions - time, moisture, grind size, geometry - can influence select chemical groups, enabling the optimization of the process in tailoring an instant coffee aroma profile through process engineering.

582249 Quantifying the Extent and Rate of Caking in Food Powders

"Caking" is the phenomenon where materials gain strength during storage due to changes in the particle/particle interactions. A significant number of materials, including raw feeds, additives and process intermediates, from a wide range of industries are susceptible to these changes. Following storage, previously free-flowing and easily processable powders can have limited ability to pass uninterrupted through their process and this can have a detrimental impact on product quality. Therefore, the ability to quantify the extent, strength and rate at which a powder cakes is vital to process operators and designers. This presentation demonstrates how the caking of food powders can be quantified using powder flow characterization techniques.

Caking can occur via a number of different mechanisms: mechanical, thermal, environmental and/or chemical. A combination of mechanisms is discussed in this study, where the high levels of fats and sugars within food powders can also intensify the strength and rate of caking. Some caking mechanisms can be controlled by managing environmental conditions, however this can be impractical and uneconomical. Accurately quantifying unacceptable characteristics and the stimuli that contribute towards caking for particular powders is therefore highly advantageous.

A range of food powders were evaluated to investigate the effect of extended storage time and environmental conditions on their flow properties. Samples were placed into 25 ml cylindrical vessels and initially conditioned using an FT4 Powder Rheometer® (Freeman Technology Ltd, UK). Conditioning was completed by passing a bespoke, twisted blade through the powder along a prescribed helical pathway. This promoted the formation of a uniform powder bed with a repeatable stress state within the sample. The test samples were then stored in an environment at controlled humidity and temperature levels for predetermined time periods. Stored samples were analyzed using the same bespoke blade and test parameters to measure the powder's resistance to motion and quantify the relative change in flow properties as a result of caking.

Analysis of the powder's resistance to motion with respect to blade position provides strong evidence of localized caking within some materials. At elevated humidity levels, some powders gained strength at the powder-air interface, forming a "crust". It was possible to easily evaluate both the strength and depth of the crust and improve understanding of moisture migration through the powder bed.

This study demonstrates that caking is a complex phenomenon and that powder flow characterization is a relevant technique to quantify the progression of caking as a function of time, relative humidity and temperature. In turn, this characterization can assist with adapting the process environment or parameters to limit caking and retain optimal processability.

563384 Optimization of Cleaning in Process Equipment Via CFD Studies

In the fine chemical, pharmaceutical, and food industries, a variety of methods are used to clean work surfaces between production runs. To reduce downtime and the quantities of cleaning agents that are used, there is a need to ensure that cleaning operations are completed quickly and efficiently. Common processes for cleaning vessels and industrial equipment include soaking, reflux, and washing with a spray or jet of a cleaning agent. To clean the inner surfaces of piping systems, fluids may be pumped through them. The choice of cleaning method largely depends on the nature of the residues and the mechanisms by which they can be removed. For example, residues may be removed by either mechanical force (shear stress) or dissolution into a flowing cleaning agent, often a solvent or detergent solution.

We report on the use of computational fluid dynamics (CFD) to model four examples of cleaning processes. Through simulations, we estimate the distributions of the forces exerted on surfaces and the mass transfer (dissolution) rates during soaking, reflux, impingement of a solvent jet, and pumping. All studies are three-dimensional and involve turbulent flows. For the single-phase flow systems (pumping and soaking with stirring), we use the Reynolds-averaged Navier-Stokes (RANS) equations coupled with a high Reynolds number k-ε turbulence model and standard wall function. To model an impinging jet and falling condensate film (cleaning by reflux), we use the volume of fluid (VOF) method to describe the gas-liquid flow together with large eddy simulation (LES) for turbulence modelling. In all studies, we consider the removal of a layer of acetaminophen (paracetamol) by water or methanol, examples of liquids in which acetaminophen has a low and high solubility, respectively. The dissolution and transport of acetaminophen is modelled through a passive scalar transport equation. We use acetaminophen as one example of a molecule with a representative solubility and diffusivity for the types of residues that are encountered in pharmaceutical manufacturing.

Sample computational results are shown in the accompanying figure for the studies of solvent flow in a pipe (a) and an impinging solvent jet (b). The piping system has two straight sections, with one joined to the middle of the other to model a case with a dead leg. Residue removal rates by dissolution (solute fluxes) are shown along a line on the side of the longer pipe section. Part (b) shows the solute fluxes at varying radial distances from the impingement point of a jet of water or methanol with a solid surface. In these systems, operating parameters such as temperature have several effects, for example on the solvent viscosity and residue solubility and diffusivity. In the case of a jet, the effects on the cleaning rate are complex, and CFD offers a way to estimate cleaning rates as a function of system parameters. Furthermore, CFD can be used to assess the locations of hard-to-clean areas within vessels and piping systems and evaluate options to improve cleaning. Use of CFD therefore appears to be promising for predictive modelling of cleaning processes to assist optimization and ensure successful completion (and validation).

The research leading to these results has received funding from Enterprise Ireland Innovation Partnership Programme (IP-2017-0573).

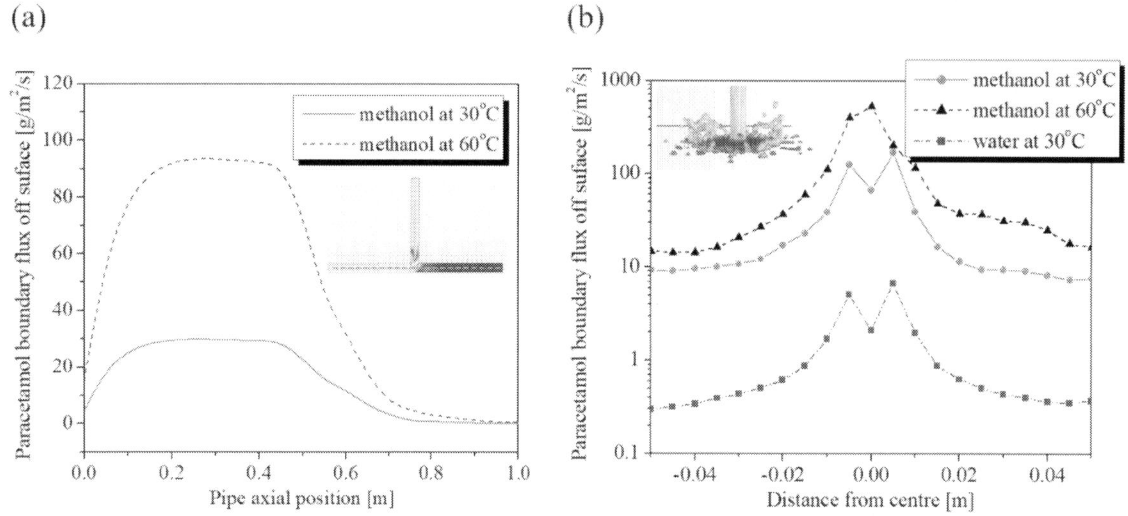

Fig. 1. Spatial distributions of contaminant flux from (a) the inner surface of a 2" diameter pipe into flowing methanol at 30 °C and 60 °C and (b) a flat wall with an impinging jet of methanol at 30 °C, 60 °C, and water at 30 °C. Inset in (a) shows the solvent velocity magnitude (2 m/s average inlet speed) in a cross-section of the pipe junction, solvent flows from the left to the top (right end is closed), and solute fluxes are evaluated along the dashed orange line. Inset in (b) shows a visualization of the shape of the solvent jet soon after impact with the surface.

568460 Development of a CFD-DEM Model in Non-Inertial Frame for Solid-Liquid Mixing Applications

Solid-liquid mixing is a significant process that is critical to pharmaceutical, biomechanical, food processing and other industries. Despite the relevance of solid-liquid mixing applications, the physics governing the suspension of particles in agitated vessels is still poorly understood. This can lead to sub-optimal designs and important energy waste or even an incapacity to achieve the desired mixing state. Important criteria that characterize solid-liquid mixing, like the just-suspended speed (N_{js}) and the homogeneity of the suspension, are difficult to predict. Consequently, engineers often rely on correlations to design solid-liquid mixing vessels. There is a need for more studies on solid-liquid mixing and, more particularly, for new and robust simulation models to predict suspension dynamics.

Complex geometries can be used to meet specific process requirements. For example, the double helical ribbon has shown to achieve good mixing in the laminar regime. However, the simulation of solid-liquid flow in such rotating systems is challenging. To deal with this issue, common techniques such as the sliding mesh and multiple reference frame can be used, with known limitations related to computational cost and accuracy .The present work concerns the development of an Euler-Lagrange model in a non-inertial frame of reference that is suitable to a wide range of geometries regardless of their complexity. It is based on an unresolved CFD-DEM approach that combines the Discrete Element Method (DEM) for the solid particle dynamics and CFD techniques for the fluid phase [1].

First, the full model is introduced. Using simple numerical experiments, the DEM sub-model in a rotating frame of reference is then assessed, including its stability, which is not simple because of the addition of a velocity-dependent force related to the Coriolis effect. Next, our coupling strategy for the fluid and the solid phases is described and the full CFD-DEM model is validated against experimental data and numerical studies previously carried in our lab in the case of simple geometries like the pitch blade turbine [2]. Finally, the model is applied to the double helical ribbon in order to evaluate the impact on the quality of mixing of factors such as the rotation speed and the fluid viscosity.

[1] B. Blais, M. Lassaigne, C. Goniva, L. Fradette, and F. Bertrand, "Development of an unresolved CFD–DEM model for the flow of viscous suspensions and its application to solid–liquid mixing," *Journal of Computational Physics,* vol. 318, pp. 201-221, 2016/08/01/ 2016.

[2] M. Lassaigne, B. Blais, L. Fradette, and F. Bertrand, "Experimental investigation of the mixing of viscous liquids and non-dilute concentrations of particles in a stirred tank," *Chemical Engineering Research and Design,* vol. 108, pp. 55-68, 2016.

557159 Design Optimization of Contra-Rotating, Baffle-Free Impellers for Maximizing Solid-Liquid Mixing Efficiency

The objective of this study is to use the Taguchi method to manipulate the parameters in the mixer design, which can reduce the number of Computational Fluid Dynamics (CFD) simulation cases, thereby decreasing the total experiment time. In this study, the contra-rotating, baffle-free impeller with the inward flow is studied with various designs of mixers derived from the orthogonal array of Taguchi method. This work is focuses on mixing operations involving solid and liquid phases. Solid particles with specific gravities of 0.866 and 1.050 $g.cm^{-3}$ are introduced into CFD simulation. The mixing efficiency with different specific gravities of solid particles is used to find the optimal design. The results show that the impeller clearance is a significant factor on the fluid flow and thus the mixing proficiency. A combination of CFD simulation and Taguchi method can be optimize and characterize the design parameters that affect mixing efficiency, which can be used as a design guideline of for this impeller type.

572149 Immiscible Liquid-Liquid Mixing Strategy in an Agitated Chemical Reactor – a Computational Fluid Dynamics Study

Immiscible liquid-liquid mixing strategy in an agitated chemical reactor – A Computational Fluid Dynamics study

Rui C. Silva, Maria Inês Lopes, Filipe Neves

Immiscible liquid-liquid mixing performed in agitated chemical reactors is a fundamental unit operation in batch pharmaceutical and chemical production. The efficiency of said unit operation is affected by a number of operational parameters such as impeller geometry, number of impellers, number of baffles, reactor geometry, power number and dispersed phase volume fraction as well as physical properties of phases such as viscosity and density.

CFD simulations are a cost-effective tool to calculate mixing efficiency particularly for complex impeller geometries. A significant number of advantages of a numerical CFD approach to processes involving complex impeller geometries, when compared with traditional experimental measurement techniques: overall enhancement of process knowledge on the influence of several operational parameters is attained for every point of the study domain in a reduced timespan, minimized burdening the process train with scale testing and easy troubleshoot.

In this work a numerical study using an open-source CFD package, OpenFOAM, was performed to characterize several impeller geometries as well as assessing their impact on the mixing efficiency of immiscible liquids. The efficiency of mixing (i. e. mixing index) was assessed using the numerical velocity field, energy consumption and the amount of dispersed phase transferred. Different impeller geometries, number of baffles and immiscible liquids addition strategy were tested using Computational Fluid Dynamics (CFD). Steady state and transient flow fields were calculated for given flow conditions using solvers based on simpleFoam (Semi-Implicit Method for Pressure-Linked Equations) and pimpleFoam (merged PISO-SIMPLE) respectively. Turbulence was modelled by the two-equation standard high-Re k-epsilon turbulence model, as implemented in OpenFOAM, in tandem with a Moving Reference Frame (MRF).

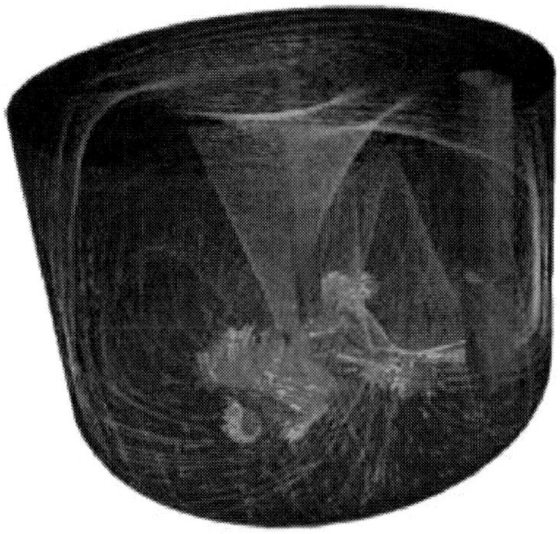

Figure – Velocity vector and streamlines on bottom impeller with one baffle.

578482 The Effect of the Presence of Internals on the Flow Regime in Industrial-Scale Bubble Column Reactor

In this work, for the first time, the effects of the variation in the low aspect ratio and the presence of heat exchanging internals on the flow regime transition in an industrial-scale pilot- plant bubble column have been investigated. The flow regime transition has been demarcated experimentally by the linear method (overall gas holdup and drift flux), and the non-linear method (Kolmogorov Entropy (KE)) using the liquid extension technique and the differential pressure technique, respectively. While, an advanced four-point optical fiber probe has been utilized to demarcate the transition velocities and characterize the bubble properties, which are the local gas holdup, bubble chord length, and bubble passing frequency, of each flow regime. The experiments were executed in an industrial-scale bubble column of I.D. 0.6 m and 3.96 m height that occupied by a heat exchanging internals covering 24 % of the total cross-sectional area of the bubble column. The superficial gas velocity varied from 0.005 m/s to 0.45 m/s.

579222 On the Joint Use of Simulation and Experimentation to Clarify Solid-Liquid Mixing Behavior

Solid-liquid mixing processes are commonly used in a wide range of industries for the production of consumer goods such as pastes, paints, cosmetics, pharmaceuticals and food products. Despite this importance, the determination of optimal process operating conditions that lead to the desired properties for the suspensions to be produced remains difficult task. To shed light on this issue related to solid-liquid mixing, simulation and experimental work is essential.

Many examples are presented here to illustrate what has been done in our research group in recent years to better understand solid-liquid mixing through the development of simulation models and pressure-based experimental techniques. These models are based on a so-called unresolved CFDDEM strategy that combines CFD techniques for the fluid flow and the discrete element method (DEM) for the particle dynamics. The experimental techniques are extensions of the pressure gauge technique, whereby pressure variations are measured at different heights throughout the tank by the use of pressure sensors. While the focus is on non-dilute solid-liquid mixing in a stirred tank provided with a pitched-blade turbine in the laminar, transitional and turbulent regimes, results obtained in the case of close-clearance impellers are also discussed.

AUTHOR INDEX

AUTHOR INDEX